Please return to
G. W. Patterson Jr.
#88 W. N.

A TEXT-BOOK

OF

ELEMENTARY CHEMISTRY

THEORETICAL AND INORGANIC

By GEO. F. BARKER, M. D.
PROFESSOR OF PHYSIOLOGICAL CHEMISTRY IN YALE COLLEGE.

LOUISVILLE, KY.:
JOHN P. MORTON AND COMPANY,
PUBLISHERS.

PREFACE.

Within the past ten years, Chemical science has undergone a remarkable revolution. The changes which have so entirely altered the aspect of the science, however, are not, as some seem to suppose, changes merely in the names and formulas of chemical compounds; for in this, the science is but returning to principles long ago established by Berzelius. They are changes which have had their origin in the discovery, first, that each element has a fixed and definite combining power or equivalence; and second, that in a chemical compound, the arrangement of the atoms is of quite as much importance as their kind or number. The division of the elements into groups, according to the law of equivalents, necessitated a revision, and in some cases, an alteration, of their atomic weights; while, in obedience to the second law, molecular formulas were reconstructed so as to express this atomic arrangement. The importance of these laws can not be overestimated. By the former, all the compounds formed by any element may be with certainty predicted; by the latter, all the modes of atomic grouping may be foreseen, and the possible isomers of any substance be pre-determined. Instead, therefore, of being a heterogeneous collection of facts, Chemistry has now become a true science, based upon a sound philosophy.

The first part of this book is intended to be an elementary treatise upon Theoretical Chemistry. It aims to present the principles of the science as they are held by the best chemists of the day upon a new plan of treatment which the author has found simple and satisfactory in his own teaching. In studying it, it is desirable that the student commit to memory the portions given in large type; while the examples given in small type, he may recite in his own language. These, it must be remembered, are to be extended by the teacher until the principles they illustrate are clear to every mind. The questions and exercises printed at the end of each chapter, are intended to be suggestive rather than exhaustive; these, therefore, should be amplified by the teacher at his discretion. By means of the table on page 16, the class should be thoroughly drilled in the rules of naming chemical compounds; and by this table, used in connection with those on pages 14 and 19, a very thorough drill in chemical notation may be secured.

The second part of the book contains the facts of Inorganic Chemistry, arranged systematically under appropriate heads. To as great an extent as seemed desirable, theory has been applied to explain the formation and properties of compounds. The unsatisfactory classification of the elements into metals and metalloids is discarded and they are arranged electro-chemically, from negative to positive. The problems given in the exercises should be conscientiously worked out by the student. The metric system of weights and measures, and the centigrade scale of thermometric degrees is used throughout the book. Tables in the Appendix show the relation of these measurements to those of our ordinary standards.

The entire book, it is believed, is a fair representation of the present state of Chemical science. If much appears in it that is novel, much more has been omitted because unsuited to a strictly elementary book. The author has made free use of all the works accessible to him, both for his facts and for many of his illustrations. For the former, he is indebted to Frankland's "Lecture-notes"—from which the definition of Chemistry is almost literally taken; to the "Grundlehren der Theoretischen Chemie" of Buff, the "Chemie der Jetztzeit" of Blomstrand, and the "Lehrbuch der Chemie" of Geuther; as well as to a paper on the "Method of teaching advanced classes in Chemistry" by Professor Cooley of Albany. For the latter, to the "Lehrbuch der Anorganischen Chemie" of Arendt, to the "Leçons élémentaires de Chimie moderne" of Ad. Wurtz, and to the manual of Chemistry by Bloxam.

In conclusion, this text-book is offered as a contribution toward making science disciplinary as well as instructive. If it be true that Chemistry already excels in training the powers of perception and of memory, it is unquestionably true that this science is capable of developing the reasoning faculties also. The present attempt to make it available for this purpose therefore, may fairly ask to be judged, not in the light of its shortcomings alone, but also by the desirability of the end at which it aims.

NEW HAVEN, October, 1870.

TABLE OF CONTENTS.

PART FIRST.—THEORETICAL CHEMISTRY.

PART SECOND.—INORGANIC CHEMISTRY.

PART FIRST.

Theoretical Chemistry.

Part First.

THEORETICAL CHEMISTRY.

CHAPTER FIRST.

INTRODUCTION.

§ 1. Physical and Chemical Properties of Matter.

1. Natural and Physical Science.—Science is a classified knowledge of external nature. It is divided into Natural and Physical Science.

Natural Science considers the external form and the internal structure of bodies.

Physical Science concerns itself only with the matter of which these bodies are made up.

Examples.—Geology, Mineralogy, Botany, and Zoölogy, which treat of the form and structure of the earth, of minerals, of plants, and of animals, respectively, are Natural Sciences.

Physics and Chemistry, which study the properties of matter itself, whether hard or soft, heavy or light, combustible or non-combustible, are Physical Sciences.

2. In studying Matter, Physical Science considers:—

1st. The divisions of which it is capable.

2d. The attractions by which the divided particles are held together.

3d. The motions which these particles may have.

3. Divisions of Matter.—Three divisions of matter are recognized in science: masses, molecules, and atoms.

A Mass of matter is any portion of matter appreciable by the senses.

A Molecule is the smallest particle of matter into which a body can be divided without losing its identity.

An Atom is the still smaller particle produced by the division of a molecule.

EXAMPLES.—The sun and the sand-grain are equally *masses* of matter. The smallest particles of sugar or of salt which can exhibit the properties of these substances, respectively, are *molecules* of sugar or of salt. The still more minute particles of carbon, of hydrogen, and of oxygen, which make up the molecule of sugar, or those of chlorine and of sodium which compose the molecule of salt, are *atoms*.

A mass of matter is made up of molecules, and a molecule itself is composed of atoms.

4. Attractions of Matter.—The forms of attraction admitted in science are three in number:—

1st. That form of attraction which is exerted between masses of matter, called gravitation.

2d. That which binds molecules together, called cohesion, when the molecules are alike; adhesion, when the molecules are unlike.

3d. That which takes place between atoms, called chemism.

EXAMPLES.—The planets are held to the sun, and the pebble is held to the earth, by the attraction of *gravitation*. The molecules of gold or of salt, being alike, are held together by *cohesive* attraction; those of granite or of gunpowder, being unlike, are held together by *adhesion*. The atoms which make up a molecule of gold or of salt, are united by *chemical* attraction.

Like molecules united by cohesion, form homogeneous matter; unlike molecules, united by adhesion, form heterogeneous matter. There are as many kinds of molecules as there are kinds of homogeneous matter.

5. Motions of Matter.—Three kinds of motion are recognized in science:—

1st. Mass-motion, or visible mechanical motion.

2d. Molecular motion, the motion of the molecule within the mass, called Heat, Light, Electricity or Magnetism, according to its character.

3d. Atomic motion; which, though probable, is not yet fully established.

6. Province of Physics.—Physics is that department of Physical Science which studies the results which flow from the molar and molecular conditions of matter.

EXAMPLES.—*Weight*, which is a consequence of mass-attraction, and *impact*, which is a result of mass-motion; *tenacity*, *hardness*, and *elasticity*, which depend upon cohesion, and *solution*, *capillarity*, and *diffusion*, which result from adhesion; and the phenomena of *heat*, *light*, *electricity* and *magnetism*, which are molecular motions, are all objects of Physical study and investigation.

7. Province of Chemistry.—Chemistry, on the other hand, studies matter in its atomic condition. Its province is to account for the differences observed in the various kinds of homogeneous matter, and to investigate the changes which this matter may undergo.

8. Physical and Chemical changes.—Physical changes in matter are those which take place outside the molecule; they do not affect the molecule itself and therefore do not alter the identity of the matter operated on.

Chemical changes take place within the molecule; they alter the character of the molecule, and hence cause a change in the matter itself.

EXAMPLES.—Water, which is a liquid, may change to ice, a solid, or become steam, a gas. But in all these forms, the molecule is identically the same; these changes are therefore *physical.* But when water is subjected to the influence of electricity, it undergoes a more radical change; the water disappears, and in its place appear two gaseous substances, oxygen and hydrogen, with entirely unlike molecules, and hence with entirely different properties from the water in either of its physical states. Such a change as this is a *chemical* change.

Any change in matter which destroys the identity of the substance acted on, is a chemical change. All others are physical changes.

9. Physical and Chemical Properties.—Physical properties are those which bodies possess in virtue of their molar or molecular condition. Chemical properties are those which are a result of the atomic composition of the molecule.

EXAMPLES.—*Density,* the attraction of the earth's mass for the unit of volume of any body; *tenacity,* which measures the strength of cohesive attraction; *color,* a result of the action of the molecules upon light; *physical state,* depending on the character of the molecular motion; these are examples of *physical* properties.

Combustibility, explosibility, capability of union with, and *of action upon, other bodies,* are examples of *chemical* properties.

10. Differences in Molecules.—Molecules differ from each other because the atoms of which they are composed differ. This difference may be,—

1st. In kind.

2d. In number.

3d. In relative position.

EXAMPLES.—A molecule of salt, made up of atoms of chlorine and sodium, differs from a molecule of water, composed of oxygen and hydrogen atoms, because the *kind* of atoms in each molecule is different. A molecule of corrosive sublimate and one of calomel are very distinct bodies, though both are composed of atoms of mercury and chlorine; the former containing only half as many mercury atoms as the latter, to the same weight of chlorine; in this case the difference is due to the *number* of the component atoms. Starch, woody fiber, and gum, substances obviously different, contain in their molecules, not only carbon, oxygen and hydrogen atoms, but precisely the same number of each ; the differences in these bodies, therefore, can be due only to a difference in the *relative position* of these atoms in the molecule of each substance.

11. Chemistry Defined. — Chemistry is that branch of Physical Science which treats of the atomic composition of bodies, and of those changes in matter which result from an alteration in the kind, the number, or the relative position of the atoms which compose the molecule.

12. Mutual Relations of the Sciences.—All branches of science are mutually dependent. The Natural Sciences require the aid of Physics and Chemistry to explain many of their phenomena ; and Physical Science finds some of its best illustrations in Mineralogy and Zoölogy. The physical properties of a body have a most intimate connection with its chemical composition ; and chemical changes are modified to a very large extent by physical conditions. The student of Physical science must know something, therefore, of the laws of form and structure ; and the chemical investigator, to be successful, must have an accurate knowledge of physical principles.

TABULAR RECAPITULATION.

Sciences.	*Divisions.*	*Attractions.*	*Motions.*
Physics	Masses	Gravitation	Mechanical power.
	Molecules	Cohesion Adhesion	Heat. Light. Electricity. Magnetism.
Chemistry	Atoms	Chemism	———

EXERCISES.

1. What is Science?
2. How is Natural distinguished from Physical Science?
3. Under which head would Astronomy be classed?
4. Is Physiology a Natural or a Physical Science?
5. What are the divisions of matter?
6. Define a molecule of water.
7. Of what is a mass constituted? A molecule?
8. Define gravitation, cohesion, chemism.
9. What is the difference between homogeneous and heterogeneous matter?
10. Define mechanical motion, heat, electricity.
11. Are the Atoms of bodies in motion?
12. What are the objects of Physical study? Illustrate.
13. What is the province of Chemistry?
14. Is the explosion of gunpowder a physical or a chemical change?
15. How are Physical properties distinguished from Chemical?
16. Why is Weight a physical property?
17. Why is Combustion a chemical phenomenon?
18. In what ways may molecules differ from each other?
19. Why do water and calomel differ from each other?
20. Give the definition of Chemistry.
21. Is any department of Science absolutely independent?

CHAPTER SECOND.

ELEMENTAL MOLECULES AND ATOMS.

§ 1. MOLECULES IN GENERAL.

13. Chemical Definition of Molecule.—A molecule is a group of atoms united by chemism. It is the smallest particle of any substance which can exist in a free or uncombined state in nature.

14. Analysis and Synthesis.—The chemist ascertains the composition of molecules by two methods :—

1st. By analysis, which consists in separating the molecule into its constituent atoms.

2d. By synthesis, which consists in putting together the constituent atoms to form the molecule.

15. Classification of Molecules.—Molecules are divided into two classes :—

1st. Elemental molecules, in which the atoms are alike.

2d. Compound molecules, in which the atoms are unlike.

Matter made up of molecules containing like atoms, is called simple or elementary matter; matter whose molecules are made up of dissimilar atoms, is called compound matter.

EXAMPLES.—A mass of iron, of charcoal, or of sulphur, is formed of molecules whose atoms are alike; these substances therefore, are examples of *simple* or *elementary matter*. The molecules which compose a mass of glass, of marble, or of water, are made up of dissimilar atoms ; these substances are examples of *compound matter*. To the latter class, by far the larger number of substances belong.

§ 2. Elemental Molecules.

16. Mode of distinguishing Elemental from Compound Molecules.—Elemental molecules may be distinguished from those which are compound, by causing a re-arrangement of the atoms between two similar molecules. Elemental molecules under this treatment, yield no new kinds of matter ; compound molecules yield elemental molecules, which, of course, are different in kind.

Examples.—Let *aa* and *aa* be two molecules, each composed of the similar atoms *a* ; then by re-arrangement, *aa* and *aa* will be produced, differing not at all from the original molecules. If however, the molecules be *ab* and *ab*, composed of dissimilar atoms, then by re-arrangement, *aa* and *bb* are produced, both elemental molecules. So *ac* and *ac* would give *aa* and *cc* ; *bd* and *bd*, *bb* and *dd*.

From simple matter only simple matter comes ; but from compound matter, simple matter is obtained.

Silver, oxygen, copper, yield in this way only silver, oxygen, copper. But salt gives chlorine and sodium, water gives oxygen and hydrogen, blende gives zinc and sulphur by this re-arrangement.

17. Number of Elemental Molecules.—By effecting this re-arrangement with the molecules of all the various kinds of homogeneous matter known, the number of different elemental molecules has been ascertained. This number is sixty-three. Moreover, since every distinct elemental molecule is made up of atoms peculiar to itself, it is obvious that the number of different kinds of atoms must be sixty-three also.

This is the number at present known. Should some new molecule of compound matter be examined hereafter, or should some molecules not now supposed compound, be proved to be so, others might be added to the list.

Of these sixty-three kinds of atoms, all molecules, and therefore, all masses of matter, are made up.

18. Names of Elemental Molecules and Atoms.—Both the elemental molecules and their constituent atoms have the same name. In many cases, this name is the one by which the simple substance is familiarly known. But in general, the name is given by the discoverer, either in allusion to some special property of the body, or to its origin or association. In a few instances the names are fanciful.

EXAMPLES.—The name *gold* is applied equally to the molecules which make up the mass, and to the atoms which form the molecule. This name, like that of lead, tin, or iron, is the name of the substance in common use.

Chlorine, a greenish-yellow gas, gets its name from χλωρός, greenish-yellow. Hydrogen comes from ὕδωρ water, and γεννάω I produce, because it generates water by its combustion. Calcium comes from *calx*, lime, because it is obtained from that substance. Cæsium is from *cæsius*, sky-blue, because there are two bright blue lines in its spectrum. Cerium, Palladium, and Uranium are named from the corresponding planets. And Titanium and Tantalum from mythological deities.

(A list of the sixty-three simple substances, with the derivation of their names, and the names of their discoverers, is given in the appendix.)

19. Size and Weight of Elemental Molecules. — According to the law of Ampère, equal volumes of all bodies in the gaseous state, contain the same number of molecules. From which it follows :—

1st. That the molecules of all bodies when in the gaseous state, are of the same size.

2d. That the weight of any molecule—compared with that of hydrogen—is proportional to the weight of any given volume—also compared with the same volume of hydrogen.

EXAMPLES.—If one liter of oxygen—which weighs 16 times as much as a liter of hydrogen—contains the same number of molecules, then it is obvious that each molecule of oxygen must be 16 times as heavy as a molecule of hydrogen. If the density of nitrogen be 14, a molecule of nitrogen must be 14 times heavier than a molecule of hydrogen.

20. Number of Atoms in the Hydrogen Molecule.—Assuming that one volume of hydrogen contains 1000 molecules, then by Ampère's law, one volume of chlorine will contain 1000 molecules also. If these volumes be mixed together and exposed to sunlight, they combine to form two volumes of a new substance, hydrochloric acid gas, which two volumes, of course, by the same law, will contain 2000 molecules. Upon analysis, each molecule of hydrochloric acid gas is found to consist of two atoms, one of hydrogen, the other of chlorine. The 2000 molecules, therefore, will contain 2000 hydrogen-atoms and 2000 chlorine-atoms. But the 2000 hydrogen-atoms came from the 1000 molecules in the original volume; and the 2000 chlorine-atoms came from the 1000 chlorine molecules. Each molecule of hydrogen must therefore have furnished two hydrogen-atoms; and each molecule of chlorine, two chlorine-atoms. Hence a molecule of hydrogen is made up of two atoms.

21. Molecular Weights.—If the weight of the hydrogen atom be taken as 1, then—since its molecule contains two atoms—its molecular weight will be 2. The molecular weight of any other substance may be obtained by multiplying its density in the state of gas, by the molecular weight of hydrogen.

EXAMPLES.—The density of nitrogen gas, for example, is 14; that is, a given volume of it—say one liter—weighs 14 times as much as one liter of hydrogen. Its molecule must therefore be 14 times as heavy; and since the molecular weight of hydrogen is 2, the molecular weight of nitrogen is 14×2; that is, 28. Again, the density of phosphorus vapor is 62. Its molecular weight is 62×2, or 124.

22. Number of Atoms in Elemental Molecules.—The number of atoms in any elemental molecule is obtained by dividing the molecular weight by the atomic weight.

EXAMPLES.—The molecular weight of nitrogen being 28, and its atomic weight—obtained by a method to be described hereafter—14, the number of atoms in its molecule is $28 \div 14$, or 2. The molecular weight of phosphorus being 124, and its atomic weight 31, its molecule must contain four atoms.

23. Atomicity of Elemental Molecules.—The atomicity of a molecule is the number of atoms which it contains. Molecules are mon-atomic, di-atomic, tri-atomic, tetr-atomic, or hex-atomic, according as they contain one, two, three, four, or six atoms. Most elemental molecules are di-atomic.

(Many simple substances, not being volatile, cannot be weighed in the gaseous state. They are usually classed as di-atomic, however, their di-atomicity being assumed.)

Those elemental molecules whose atomicity has been experimentally determined, are given in the following table:—

Mon-atomic.	*Di-atomic.*	*Tri-atomic.*	*Tetr-atomic.*	*Hex-atomic.*
Mercury	Hydrogen	Oxygen (as ozone)	Phosphorus	Sulphur
Cadmium	Oxygen		Arsenic	
Zinc	Chlorine			
Barium	Bromine			
	Iodine			
	Fluorine			
	Nitrogen			
	Sulphur			
	Selenium			

§ 3. PROPERTIES OF ATOMS.

24. Definition of Atom.—An atom is the smallest particle of simple matter which can enter into the composition of a molecule.

25. Classification of Atoms.—Atoms differ from each other :—

1st. In weight.

2d. In the quality of their combining power.

3d. In the quantity of their combining power.

26. Atomic Weight.—The relative weight of any atom referred to hydrogen as unity, is its atomic weight. It is the smallest quantity of any simple substance, by weight, which can take part in the formation of any chemical compound.

27. Method of fixing the Atomic Weight.—In order to fix an atomic weight, we must know :—

1st. The quantity by weight of the substance which combines with one atom of hydrogen.

2d. The molecular weight of the hydrogen compound.

I. Analysis of the compound which any body forms with hydrogen, will give the absolute weight of each constituent in any given quantity—say 100 parts. The quantity of the body by weight which unites with one part of hydrogen may then be obtained by a simple proportion.

EXAMPLES.—Analysis shows that hydrochloric acid gas—the hydrogen compound of chlorine—contains in one hundred parts, 97·26 parts of chlorine and 2·74 parts of hydrogen. By the proportion 2·74 : 97·26 : : 1 : 35·5, it is ascertained that in this compound, the quantity of chlorine which combines with one part, i. e., one atom, of hydrogen, weighs 35·5 times as much.

Again the analysis of water shows it to be composed of 88·89 per cent of oxygen and 11·11 per cent of hydrogen; whence 11·11 :

88·89 : : 1 : 8. Eight parts of oxygen unite with one part of hydrogen.

So ammonia gas contains in 100 parts, 82·35 parts of nitrogen and 17·65 parts of hydrogen; or 17·65 : 82·35 : : 1 : 4·7. Hence 4·7 parts of nitrogen combine with 1 part of hydrogen.

Now, if in a molecule of hydrochloric acid, water, or ammonia, there is but one atom of hydrogen, then 35·5, 8, and 4·7, being the smallest quantities by weight in which chlorine, oxygen, and nitrogen combine, will be the atomic weights of these bodies respectively.

II. The molecular weight of any substance is the sum of the weights of its constituent atoms. The combining weights obtained by analysis, when added together, therefore, must either give the molecular weight, or some number of which the molecular weight is a multiple. In this way the number of hydrogen atoms in the compound may be determined; and the quantity by weight of the other constituent which is united with these hydrogen atoms, is its atomic weight.

EXAMPLES.—The density of hydrochloric acid gas, of steam, and of ammonia gas respectively, is 18·25, 9, and 8·5; their molecular weights therefore, are 36·5, 18 and 17. The sum of the combining weights of hydrogen and chlorine obtained by analysis, (35·5+1) is 36·5: but this is the molecular weight of hydrochloric acid gas. Hence this gas contains one atom of chlorine and one of hydrogen, and the atomic weight of chlorine is 35·5.

Again, the sum of the combining weights of oxygen and hydrogen, (8+1) is 9. But 9 is one half the molecular weight of water; hence, a molecule of water contains twice as much of each constituent; i. e., 2 parts of hydrogen and 16 of oxygen. 16 is therefore the atomic weight of oxygen.

So, if the combining weights of nitrogen and hydrogen be added together, the sum is (4·7+1) 5·7. But 5·7 is only one-third the molecular weight of ammonia, which is 17. Ammonia therefore, contains three times as much hydrogen and nitrogen in one molecule, as that given by analysis. It must contain therefore, 3 parts of hydrogen and 14 (4·7×3) parts of nitrogen; and the atomic weight of nitrogen is fixed at 14.

SYMBOLS AND ATOMIC WEIGHTS.

	Symbol.	*At. wt.*
Hydrogen,	H	1
Fluorine,	F	19
Chlorine,	Cl	35·5
Bromine,	Br	80
Iodine,	I	127
Oxygen,	O	16
Sulphur,	S	32
Selenium,	Se	79
Tellurium,	Te	128
Nitrogen,	N	14
Phosphorus,	P	31
Arsenic,	As	75
Antimony (*Stibium*),	Sb	122
Bismuth,	Bi	210
Tantalum,	Ta	182
Columbium,	Cb	94
Vanadium,	V	51·3
Boron,	B	11
Carbon,	C	12
Silicon,	Si	28
Titanium,	Ti	50
Tin (*Stannum*),	Sn	118
Chromium,	Cr	52·5
Manganese,	Mn	55
Iron (*Ferrum*),	Fe	56
Nickel,	Ni	59
Cobalt,	Co	59
Uranium,	U	120
Molybdenum,	Mo	96
Tungsten (*Wolfram*),	W	184
Osmium,	Os	199
Iridium	Ir	197

	Symbol.	*At. wt*
Platinum,	Pt	197
Palladium,	Pd	106·5
Gold (*Aurum*),	Au	196·6
Rhodium,	Ro	104·3
Ruthenium,	Ru	104·2
Mercury, (*Hydrargyrum*).	Hg	200
Copper (*Cuprum*),	Cu	63·5
Lead (*Plumbum*),	Pb	207
Silver, (*Argentum*),	Ag	108
Thallium,	Tl	204
Indium,	In	74
Didymium,	D	96
Cerium,	Ce	92
Lanthanum,	La	92
Yttrium,	Y	61·7
Erbium,	E	112·6
Glucinum,	G	9·3
Cadmium,	Cd	112
Zinc,	Zn	65
Magnesium,	Mg	24
Thorium,	Th	115·7
Zirconium,	Zr	89·5
Aluminum,	Al	27·5
Calcium,	Ca	40
Strontium,	Sr	87·5
Barium,	Ba	137
Lithium,	Li	7
Sodium (*Natrium*),	Na	23
Potassium (*Kalium*),	K	39
Rubidium,	Rb	85
Cæsium,	Cs	133

28. Indirect Method of Fixing Atomic Weights.—In some cases, elements do not unite directly with hydrogen. The comparison with hydrogen is then made by means of a middle term, generally chlorine.

EXAMPLES.—No hydrogen compound of silver is known. But silver combines with chlorine to form the well-known ore, horn-silver. On analysis this horn-silver yields 75·26 per cent of silver and 24·74 per cent of chlorine. As 35·5 parts of chlorine combine with 1 of hydrogen, the quantity of silver which combines with 35·5 parts of chlorine, is its atomic weight; 24·74 : 75·26 : : 35·5 : 108, the atomic weight of silver.

(The atomic weights of the elements are given on the opposite page.)

29. Quality of Atomic Combining Power.—Atoms are divided into two classes according to the quality of their combining power. These are:—

1st. Positive atoms, or those which are attracted to the negative pole in electrolysis, and whose hydrates are bases.

2d. Negative atoms, or those which go to the positive electrode and whose hydrates are acids.

EXAMPLES.—Salt, under the influence of electricity, is decomposed into chlorine and sodium. The chlorine atoms collect at the positive pole and are therefore called *negative*. The sodium atoms are found at the negative pole, and are hence called *positive*. Again, all the hydrates of chlorine are *acids*, while the hydrates of potassium—a substance like chlorine in all other chemical characters—are entirely different bodies called *bases*.

In the table on the next page the elements are arranged according to their electro-chemical characters, each atom being positive to any atom which is placed above it, and negative to any one given below it. These distinctions are therefore entirely relative.

Negative End —.

Oxygen.
Sulphur.
Nitrogen.
Fluorine.
Chlorine.
Bromine.
Iodine.
Selenium
Phosphorus.
Arsenic.
Chromium.
Vanadium.
Molybdenum
Tungsten.
Boron.
Carbon.
Antimony.
Tellurium.
Tantalum.
Columbium.
Titanium.
Silicon.
Hydrogen.
Gold.
Osmium.
Iridium.
Platinum.
Rhodium.
Ruthenium.
Palladium.
Mercury.
Silver.
Copper.
Uranium.
Bismuth.
Tin.
Indium.
Lead.
Cadmium.
Thallium.
Cobalt.
Nickel.
Iron.
Zinc.
Manganese.
Lanthanum.
Didymium.
Cerium.
Thorium.
Zirconium.
Aluminum.
Erbium.
Yttrium.
Glucinum.
Magnesium.
Calcium.
Strontium.
Barium.
Lithium.
Sodium.
Potassium.
Rubidium.
Cæsium.

Positive End +.

30. Quantity of Atomic Combining Power.—If the combining power of the atom of hydrogen be taken as 1, the combining power of other atoms will be 1, 2, 3, 4, 5, 6 or 7. That is, some atoms have a combining power equal to that of hydrogen and can unite with one atom of it. Other atoms have a higher combining power and can unite with 2, 3, 4, 5, 6 or 7 hydrogen atoms or their equivalents.

EXAMPLES.—Atoms combine with other atoms in virtue of their chemism, an attraction mutually satisfied by the union. Taking the chemism of an atom of hydrogen as the unit, any other atom whose chemism is completely satisfied by uniting with the hydrogen-atom, is its equal in combining power. Other atoms there are, whose chemism requires for saturation, two, three, four, five, six, or seven hydrogen-atoms ; hence their combining power—which is the same for all other similar atoms—is said to be one, two, three, four, five, six, seven.

A chlorine-atom is completely satisfied with one atom of hydrogen; but an oxygen-atom requires two, a nitrogen-atom, three, a carbon-atom, four, and so on.

31. Equivalence.—The equivalence of an atom is the quantity of its combining power, expressed in hydrogen units. It expresses the number of hydrogen-atoms it can combine with or be exchanged for.

EXAMPLES.—The equivalence of carbon is four, because one atom requires four atoms of hydrogen to satisfy its chemism. The equivalence of phosphorus is five, because it forms a compound in which one atom combines with five of chlorine. The equivalence of sulphur is two because it can replace two atoms of hydrogen.

32. Classification of Atoms, according to their Equivalence.—Atoms are called monads, dyads, triads, tetrads, pentads, hexads, and heptads,—names derived from the Greek numerals—according to their equivalence. For the adjective terms, the Latin numerals are used; atoms are univalent, bivalent, trivalent, quadrivalent, quinquivalent, sexivalent, and septivalent.

Atoms whose equivalence is even, are called artiads; those whose equivalence is odd, are termed perissads.

33. Variation in Equivalence.—Since an atom may form several compounds with the same substance, its equivalence is not invariable. It always increases or diminishes by two; so that an atom of the same element may in different compounds have an equivalence of 1, 3, 5, or 7, or of 2, 4, or 6. A perissad atom can never become an artiad atom by such a change, nor can an artiad atom become a perissad.

EXAMPLES.—Iron in green vitriol is a dyad, in pyrite it is a tetrad, and in ferric acid it is a hexad. Chlorine forms a series of compounds with oxygen, in which its equivalence is one, three, five, and seven.

The compounds formed by an atom with one equivalence are widely different in properties from those formed by the same body with a different equivalence. Sometimes the identity of the atom can be established only by the conversion of the one into the other.

The word atomicity is sometimes used to indicate the highest known equivalence of an atom.

(A table of the equivalences of atoms is given on the opposite page.)

§ 4. ATOMIC NOTATION.

34. Atomic Symbols.—In 1815, Berzelius proposed an abbreviated form of chemical language. In this system, each atom has for its symbol, the first letter of its Latin name. Where the names of two different atoms begin with the same letter, a second letter, suggestive of the name, is added.

PERISSADS.

Monads :—

Element	Valences
Hydrogen	
Fluorine	
Chlorine	I, III, V, VII.
Bromine	I, III, V, VII.
Iodine	I, III, V, VII.
Lithium	
Sodium	I, III.
Potassium	I, III, V.
Rubidium	
Cæsium	
Silver	I, III.
Thallium	I, III.

Triads :—

Element	Valences
Nitrogen	I, III, V.
Phosphorus	I, III, V.
Arsenic	I, III, V.
Antimony	III, V.
Bismuth	III, V.
Boron	
Gold	I, III.

Pentads :—

Element	Valences
Columbium	
Tantalum	
Vanadium	III, V.

Also $(V_2)^{VIII}$ and $(V_2)^{IV}$

ARTIADS.

Dyads :—

Element	Valences
Oxygen	
Sulphur	II, IV, VI.
Selenium	II, IV, VI.
Tellurium	II, IV, VI.
Calcium	II, IV.
Strontium	II, IV.
Barium	II, IV.
Magnesium	
Zinc	
Cadmium	
Glucinum	
Yttrium	
Cerium	
Lanthanum	
Didymium	
Erbium	
Mercury	$(Hg_2)''$, II.
Copper	$(Cu_2)''$, II.

Tetrads :—

Element	Valences
Carbon	II, IV.
Silicon	
Titanium	II, IV.
Tin	II, IV.
Thorium	
Zirconium	
Aluminum	$(Al_2)^{VI}$
Platinum	II, IV.
Palladium	II, IV.
Lead	II, IV.
Indium	

Hexads :—

Element	Valences
Molybdenum	II, IV, VI.
Tungsten	IV, VI.
Ruthenium	II, IV, VI.
Rhodium	II, IV, VI.
Iridium	II, IV, VI.
Osmium	II, IV, VI.
Chromium	II, IV, VI.
Manganese	II, IV, VI.
Iron	II, IV, VI.
Cobalt	II, IV.
Nickel	II, IV.
Uranium	II, IV.

EXAMPLES.—O stands for an atom of oxygen, I for one of iodine, K for one of potassium (kalium), Au for one of gold (aurum), Sn for one of tin (stannum). So Cl represents an atom of chlorine, C one of carbon, Ca one of calcium, Ce one of cerium, Cd one of cadmium, Co one of cobalt, Cr one of chromium, Cs one of cæsium, Cu one of copper (cuprum).

(The symbols of the elements are given in the table, page 14.)

35. Symbols Represent Atomic Weights.—Each atomic symbol stands not only for the atom in general, but represents particularly its atomic weight.

EXAMPLES.—Fe (ferrum) represents 56 weight-units of iron, Sb (stibium) 122 weight-units of antimony, Hg (hydrargyrum) 200 weight-units of mercury, Ag (argentum) 108 weight-units of silver; these being the atomic weights of these substances respectively.

36. Equivalence of Atoms, how Indicated.—The equivalence of an atom is indicated by placing Roman numerals above, or a little to the right of, the symbol. Sometimes minute-marks are used instead.

EXAMPLES.—$\overset{\text{I}}{\text{H}}$ or H′ stands for the monad hydrogen-atom; $\overset{\text{II}}{\text{S}}$ or S″ for the bivalent sulphur-atom; $\overset{\text{III}}{\text{P}}$ or P‴ for the trivalent phosphorus-atom; $\overset{\text{IV}}{\text{C}}$ or C⁗ for the quadrivalent carbon-atom; $\overset{\text{VI}}{\text{Te}}$ or Te^{VI} for the sexivalent tellurium-atom, etc.

37. Graphic Symbols.—The graphic symbol of an atom is a circle, with lines radiating from it to indicate its equivalence. These lines are called *bonds*. Below is a graphic representation of the seven classes of atoms:

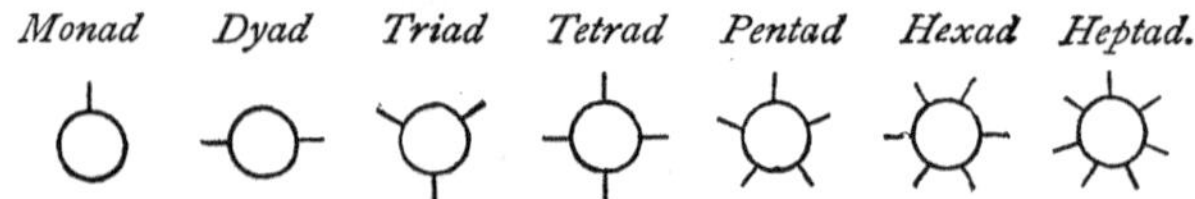

Generally however, the circles are omitted, the bonds radiating from the symbol. The *number* of bonds only is significant, not their direction.

EXAMPLES.—Thus, $-O-$, $O=$, or $\overset{|}{O}-$, stands equally for one atom of dyad oxygen. $N\equiv$, $\underset{|}{N}=$, or $-N=$, equally represents the atom of trivalent nitrogen.

38. Multiplication of Atoms. — Atoms are multiplied by placing an Arabic numeral below and to the right of the symbol.

EXAMPLES.—C_2 represents two atoms of carbon; N_4, four atoms of nitrogen; K_5, five atoms of potassium; Pt_3, three atoms of platinum.

Elemental molecules are represented in the same way:

Cl_2 stands for a molecule, composed of two atoms, of chlorine, O_3 for a molecule of ozone, As_4 for a molecule of arsenic, S_6 for a hexatomic molecule of sulphur.

Care must be taken not to confound the equivalence marks, expressed in Roman numerals, with the number of atoms, represented by Arabic numerals.

EXAMPLES.—$\overset{I}{I}_3$ stands for three atoms of iodine, each one of which is a monad. $\overset{II}{Co}_2$ stands for two atoms of bivalent cobalt; $\overset{III}{B}_5$ for five atoms of trivalent boron, $\overset{IV}{Si}_4$, for four tetrad silicon-atoms.

39. Multiplication of Molecules.—Molecules are multiplied by enclosing their symbols in brackets, and placing the numeral outside, below, and to the right.

EXAMPLES.—$(H_2)_6$ stands for six molecules of hydrogen. $(Br_2)_2$ represents two molecules of bromine. $(Na_2)_3$ indicates three molecules of sodium.

EXERCISES.

§ 1.

1. Define analysis. Synthesis.

2. What is the distinction between elementary and compound matter?

§ 2.

3. How do chemists ascertain to which class a substance belongs?

4. How many kinds of atoms are known?

5. According to what rule are the elements named?

6. Give Ampère's law and the deductions from it.

7. Show that a hydrogen molecule contains two atoms.

8. The density of arsenic-vapor is 150; what is its molecular weight?

9. The atomic weight of arsenic is 75; how many atoms are there in its molecule?

10. The density of mercury in vapor is 100; its atomic weight is 200; what is its atomicity?

11. Mention the molecules which have two atomicities.

§ 3.

12. Define an atom. An atomic weight.

13. What data are necessary to fix an atomic weight?

14. Marsh gas—a compound of hydrogen and carbon—has a density of 8; analysis shows it to be composed of 75 per cent of carbon and 25 per cent of hydrogen; what is the atomic weight of carbon?

15. The analysis of a compound of chlorine and antimony gives 46·61 per cent of chlorine and 53·39 per cent of antimony; its density is 114·25; what is the atomic weight of antimony?

16. How are positive atoms distinguished from negative?

17. Which of the following atoms are positive and which negative? Silver and carbon; tin and lead; sulphur and chlorine; sodium and iodine.

18. Define and illustrate equivalence.

19. How are atoms classified by the law of equivalence ?

20. What is a dyad ? A pentad ? A trivalent atom ? A septivalent atom ? A perissad ? An artiad ?

21. How does the equivalence of an atom vary ?

22. What equivalence has oxygen ? Nitrogen ? Iron ? Copper ?

§ 4.

23. How are atoms represented in symbols ?

24. Give the symbols of tin, zinc, silver, sodium, aluminum.

25. For what do the symbols Pb, Ca, K, W, Si, Se, Mn, Mg, stand ?

26. How are two dyad zinc-atoms written ? Four triad gold-atoms ? Six tetrad tin-atoms ? Five hexad sulphur atoms ?

27. What do $(\overset{IV}{Ti}_2)_6$, $(\overset{V}{P}_4)_2$, $(\overset{VI}{Te}_2)_4$, and $(\overset{II}{S}_6)$ represent?

CHAPTER THIRD.

COMPOUND MOLECULES.

§ 1. BINARY MOLECULES.

40. Definition.—A compound molecule is a molecule whose constituent atoms are unlike.

The number of atoms which such molecules may contain, is apparently unlimited. While a molecule of salt contains but two atoms, a molecule of protagon—a complex organic substance—contains 384.

41. Molecular Weight.—The molecular weight of a compound molecule, like that of a simple one, is the sum of the atomic weights of its constituents. It is always equal to twice the density of the substance in the state of gas.

42. Formation of Molecules.—Compound molecules are formed by the union of atoms according to the law of equivalence.

Since the number of bonds, or units of attraction, can never be less than two—one being furnished by every atom to each of the others to which it is directly united—the entire number of bonds in any molecule must be even. And since too, every perissad atom furnishes an odd number of bonds, it follows necessarily, that the number of such atoms contained in any molecule must be even.

43. Classification of Compound Molecules.—Compound molecules are divided into two classes:—

1st. Those whose atoms are directly united, called Binaries.

2d. Those whose atoms are indirectly united, called Ternaries.

44. Binary Molecules.—Binary molecules are those whose atoms are directly united. Whatever the absolute number of atoms present in such a molecule, they can never be of more than two kinds.

EXAMPLES.—A molecule of salt contains but a single atom each of chlorine and of sodium; a molecule of red-lead consists of seven atoms, three of which are lead-atoms, and four, oxygen-atoms.

45. Naming of Binary Molecules.—The names of all compound molecules are derived from those of their constituent atoms; a plan proposed by Lavoisier, in 1787.

A binary compound is formed by the union of two simple substances, one of which, by the table, must be positive to the other, which is negative. Binary molecules are named:—

By placing the name of the positive constituent first, and then the name of the negative, the termination of which is changed to ide.

EXAMPLES.

Sodium and Chlorine	yield	Sodium chloride.
Magnesium and Oxygen	"	Magnesium oxide.
Silver and Sulphur	"	Silver sulphide.
Zinc and Phosphorus	"	Zinc phosphide.
Calcium and Iodine	"	Calcium iodide.
Aluminum and Bromine	"	Aluminum bromide.
Potassium and Nitrogen	"	Potassium nitride.
Barium and Fluorine	"	Barium fluoride.
Cadmium and Selenium	"	Cadmium selenide.

The termination IDE is always characteristic of a binary compound.

46. Modification of this Rule.—Whenever the positive atom enters into combination with more than one equivalence, this fact is indicated in the compound by changing the termination of the name of this atom into ic or ous.

If the positive substance acts with but two equivalences, then in the higher one, its name takes the termination ic, and in the lower one, the termination ous.

EXAMPLES.

Bivalent Mercury	and	Oxygen	form	Mercuric oxide.
Univalent Mercury	"	Oxygen	"	Mercurous oxide.
Quadrivalent Tin	"	Chlorine	"	Stannic chloride.
Bivalent Tin	"	Chlorine	"	Stannous chloride.
Trivalent Gold	"	Iodine	"	Auric iodide.
Univalent Gold	"	Iodine	"	Aurous iodide.

If the positive constituent acts with more than two equivalences, the termination ic—being given on the discovery of the compound—is generally arbitrarily assigned. The principal groups whose equivalences thus change, with the equivalence in the ic compound, are the following:—

CHLORINE GROUP.	SULPHUR GROUP.	NITROGEN GROUP.
Equivalence V.	*Equivalence VI.*	*Equivalence V.*
Chlorine	Sulphur	Nitrogen
Bromine	Selenium	Phosphorus
Iodine	Tellurium	Arsenic

A compound in which the equivalence of the positive is less than in the ous, takes the prefix hypo, which means "under". When this equivalence is above the ic, the prefix per is given to the name of the positive. The negative termination however, in all these cases, is ide.

EXAMPLES.

Quinquivalent Chlorine	and	Oxygen	form	Chloric oxide.
Trivalent Chlorine	"	Oxygen	"	Chlorous oxide.
Univalent Chlorine	"	Oxygen	"	Hypo-chlorous oxide.
Septivalent Chlorine	"	Oxygen	"	Per-chloric oxide.
Sexivalent Sulphur	"	Oxygen	"	Sulphuric oxide.
Quadrivalent Sulphur	"	Oxygen	"	Sulphurous oxide.
Bivalent Sulphur	"	Oxygen	"	Hypo-sulphurous oxide.
Quinquivalent Nitrogen	"	Oxygen	"	Nitric oxide.
Trivalent Nitrogen	"	Oxygen	"	Nitrous oxide.
Univalent Nitrogen	"	Oxygen	"	Hypo-nitrous oxide.

47. Use of Numeral Prefixes.—In some cases the number of atoms of each constituent is to be indicated. This is done by prefixing Greek numerals to each of the names given.

EXAMPLES.

1 atom	of C	and	2 of O	form	Carbon di-oxide.
1 "	of P	"	5 of Br	"	Phosphorus penta-bromide.
2 atoms	of Fe	"	3 of O	"	Di-ferric tri-oxide.
3 "	of Ti	"	4 of N	"	Tri-titanic tetra-nitride.

48. Notation of Binary Compounds. Formulas.—Notation is the representation of a substance in symbols. If the substance be a compound one, this representation is called a formula.

Binary molecules are represented by placing the symbols of the constituent atoms together, the positive first.

EXAMPLES.

A molecule of	Hydrogen chloride	is	HCl.
" " "	Cupric oxide	"	CuO.
" " "	Silver iodide	"	AgI.
" " "	Zinc sulphide	"	ZnS.

49. Use of Numerals in Formulas.—When the number of atoms of any constituent in a molecule is more than one, this is indicated by a numeral placed below and at the right of its symbol. Compound molecules, like simple ones, are multiplied by enclosing them in brackets, and placing the numeral outside and to the right.

EXAMPLES.

SO_3	represents	a molecule	of	Sulphuric oxide.
SO_2	"	" "	"	Sulphurous oxide.
SO	"	" "	"	Hypo-sulphurous oxide.
H_2O	"	" "	"	Hydrogen oxide (water).
Pb_3O_4	"	" "	"	Tri-plumbic tetr-oxide.
$(NaCl)_2$	"	2 molecules	"	Sodium chloride.
$(CS_2)_2$	"	" "	"	Carbon di-sulphide.
$(P_2O_5)_3$	"	3 "	"	Phosphoric oxide.
$(As_2S_3)_4$	"	4 "	"	Arsenous sulphide.
(PbO_2)	"	1 molecule	"	Lead di-oxide.
$(PbO)_2$	"	2 molecules	"	Lead oxide.

50. Formation of Binaries.—Binary molecules are generally formed by the direct union of their constituent atoms. Two cases are here to be considered:—

1st. When the equivalence of the atoms is the same.

2d. When the equivalence is different.

In all cases of chemical combination, the chemism of each atom must be satisfied. Atoms having the same equivalence, then, may mutually saturate each other. Such atoms, therefore, unite in the ratio of one to one.

EXAMPLES.

K'	and Cl',	both monads,	form	$K'Cl'$,	or K—Cl.
Ca''	and O'',	both dyads,	"	$Ca''O''$,	or Ca=O.
B'''	and N''',	both triads,	"	$B'''N'''$,	or B≡N.
Pt^{IV}	and Si^{IV},	both tetrads,	"	$Pt^{IV}Si^{IV}$,	or Pt≣Si.

When the atoms which enter into combination have a different equivalence, then each atom must furnish the same number of bonds. This number is, in all cases, the least common multiple of the two equivalences. The absolute number of atoms of each constituent, is obtained by dividing this least common multiple by the equivalence of each.

EXAMPLES.—When triad- and dyad-atoms combine, each must furnish six bonds—the least common multiple of 3 and 2. But to furnish six bonds, three dyad-atoms and two triad-atoms are required. Hence the molecule will consist of two atoms of the triad and three of the dyad.

So a tetrad uniting with a dyad will require four bonds—the least common multiple of 4 and 2—to furnish which one tetrad-atom and two dyad-atoms are necessary. The following formulas still further illustrate this law:—

H', a monad, and O'', a dyad, form H'_2O''.
N''', a triad, and O'', a dyad, form $N'''_2O''_3$.
Sn^{IV}, a tetrad, and S'', a dyad, form $Sn^{IV}S''_2$.
B''', a triad, and Cl', a monad, form $B'''Cl'_3$.
Si^{IV}, a tetrad, and F', a monad, form $Si^{IV}F'_4$.
Bi^{V}, a pentad, and S'', a dyad, form $Bi^{V}_2S''_5$.
P^{V}, a pentad, and Br', a monad, form $P^{V}Br'_5$.
S^{VI}, a hexad, and O'', a dyad, form $S^{VI}O''_3$.
Cl^{VII}, a heptad, and O'', a dyad, form $Cl^{VII}_2O''_7$.

Where perissad- and artiad-atoms form a molecule, the number of atoms required is inversely as the equivalence of each.

51. Exchange of Atoms in forming Binaries.—As atoms do not exist free, they cannot in fact, form molecules by directly combining. Binary compound molecules are formed from the elemental molecules of their constituents, which, being brought into proximity, have their atoms rearranged; the chemical attractions of two dissimilar atoms being stronger than that exerted between two similar atoms.

EXAMPLES.—A molecule of magnesium $Mg=Mg$, when brought in contact with one of oxygen $O=O$, under suitable conditions, will exchange one of its magnesium-atoms for an atom of oxygen to form two molecules of magnesium oxide $Mg=O$, $O=Mg$. It may be represented thus:—

$$\text{Before}\ \begin{matrix} Mg \\ \| \\ Mg \end{matrix} \qquad \begin{matrix} O \\ \| \\ O \end{matrix} \qquad\qquad \text{After}\ \begin{matrix} Mg=O \\ \\ Mg=O \end{matrix}$$

In the same way, two molecules of hydrogen and one of oxygen will form two molecules of hydrogen oxide or water, thus :—

$$\text{Before}\quad \begin{matrix} H & & O & & H \\ | & & \| & & | \\ H & & O & & H \end{matrix} \qquad \text{After}\quad \begin{matrix} H-O-H \\ \\ H-O-H \end{matrix}$$

52. Unsaturated Molecules. Compound Radicals.—Beside the atomic groups now considered, called saturated molecules because the bonds of all the atoms they contain are mutually engaged, it is often convenient to distinguish certain unsaturated groups of atoms, which, possessing free bonds, may enter into combination like single atoms. These unsaturated groups of atoms are called compound radicals. They cannot exist in a free state in nature, though, like an atom, by combining with another similar group, they may form a molecule which is saturated. Their equivalence is always equal to the number of unsatisfied bonds ; i. e., is the difference of the equivalence of their constituents.

EXAMPLES.—Water is $H-O-H$; by removing one hydrogen atom, there is left the unsaturated group $H-O-$, which, though consisting of two atoms, is capable of entering into the formation of a molecule like any single atom. It has one free bond and hence acts as a monad. But by combining with another similar group, it forms $H-O-O-H$, a free saturated molecule. This compound radical may also be written $(H'O'')'$.

So H_3N, a saturated molecule of ammonia, yields $(H_2N)'$, a monad compound radical, by the loss of H'. $\overset{V}{P}$ a pentad phosphorus atom, may be partially saturated by one oxygen-atom O'', forming the trivalent compound radical $(P^{V}O'')'''$; or by two oxygen-atoms, giving the univalent radical $(P^{V}O_2'')'$. The former $(PO)'''$ may unite with Cl like any other triad, forming $(PO)'''Cl_3$.

53. Names of Compound Radicals.—Compound radicals have names terminating in yl. The root of the name comes either from their constituents or from some compound into which they enter.

EXAMPLES.—The compound radical $(HO)'$ is called *hydroxyl;* $(PO)'''$ is named phosphoryl; $(CO)''$ carbonyl, $(CH_3)'$ methyl, from methyl alcohol, of which it is a constituent. Three compound radicals, $(H_2N)'$ amidogen, $(CN)'$ cyanogen, and $(H_4N^V)'$ ammonium, are exceptions to this rule.

54. Artiad and Perissad Radicals.—Perissad radicals having an uneven number of free bonds, can only exist free by combining with each other, as above stated. Artiad radicals, having an even number of free bonds, may exist in the free state by the mutual saturation of these bonds by each other.

EXAMPLES.—Nitryl $(NO_2)'$, a perissad radical, cannot exist free. But by combining with another group, (NO_2)—(NO_2) is produced, which is saturated. Ethylene $\begin{array}{c} \text{H} \quad \text{H} \\ | \quad\;\; | \\ \text{H}-\text{C}-\text{C}-\text{H} \\ | \quad\;\; | \end{array}$ or $(C_2H_4)''$ is a bivalent radical; but by doubly uniting the carbon, olefiant gas $\begin{array}{c} \text{H} \quad \text{H} \\ | \quad\;\; | \\ \text{H}-\text{C}=\text{C}-\text{H} \end{array}$ results. Carbonyl $O=C=$ is a dyad radical, $O=C>$ is free carbonous oxide.

55. Explanation of Variation in Atomic Equivalence.—By a similar hypothesis, chemists explain the variation in atomic equivalence. An atom is assumed to have but one equivalence, which is the highest it ever exhibits. If now two of its bonds mutually saturate one another, the atom has a less equivalence by two; if two pairs thus saturate, by four; if three, by six. A heptad may thus become a pentad, triad, and monad successively; and a hexad, a tetrad, and a dyad; as the following graphic formulas show:—

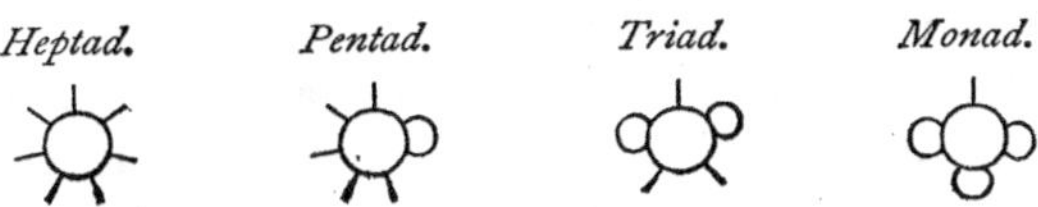

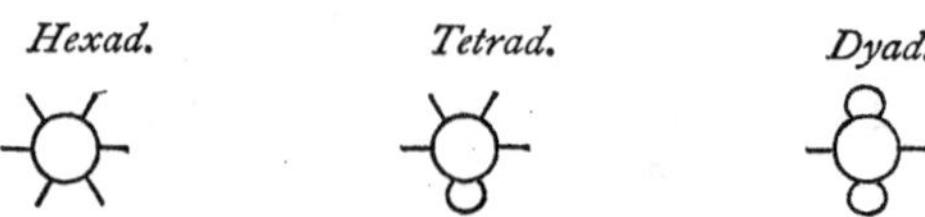

56. Names of Molecules Formed by Radicals.—Molecules which contain compound radicals united to elementary atoms are classed as binaries and are named in the same way.

EXAMPLES.—Nitryl $(NO_2)'$ and chlorine form nitryl chloride $(NO_2)'Cl$. Carbonyl $(CO)''$ and sulphur form carbonyl sulphide $(CO)''S$. Methyl $(CH_3)'$ and oxygen form $(CH_3)'_2O$, methyl oxide.

§ 2. TERNARY MOLECULES UNITED BY DYADS.

57. Classification of Ternary Molecules.—Ternary molecules are those whose dissimilar atoms are united by the aid of some third atom. This third atom, which performs a linking function, must evidently be a polyad, since no monad can join other atoms together.

Ternary molecules are divided into two classes according to the equivalence of the uniting atom :—

1st. Ternary molecules united by bivalent atoms.

2d. Ternary molecules united by trivalent atoms.

58. Ternary Molecules United by Dyads.—The dyads oxygen, sulphur, selenium, and tellurium, may perform a linking function. Of the compounds thus formed, by far the larger proportion are compounds of oxygen. Oxygen, therefore, is the distinguishing component of this class of bodies.

59. Ternary Molecules Linked by Oxygen.—Oxygen, by its two bonds, may unite two atoms or groups of atoms together. Bodies thus constituted are divided into three classes, according to the character of the atoms which are thus united. These are called acids, bases, or salts.

60. Definition of an Acid.—An acid molecule is one which consists of one or more negative atoms united by oxygen to hydrogen.

The general formula of an acid, therefore, is $\bar{R}$—O—H. The number of hydrogen-atoms which it contains is equal to the equivalence of the negative atom or group of atoms. In general, acids are recognized by the property which they possess of turning certain vegetable blues to reds.

61. Definition of a Base.—A basic molecule consists of one or more positive atoms united by oxygen to hydrogen.

A base is the analogue of an acid. It has the general formula $\overset{+}{R}$—O—H, the number of hydrogen atoms depending, as before, upon the equivalence of the positive atom or atomic group. Bases restore the color to vegetable blues which have been reddened by an acid.

62. Definition of a Salt.—A saline molecule is one which contains a positive atom or group of atoms, united by oxygen to a negative atom or group of atoms.

The general formula of a salt is $\overset{+}{R}$—O—$\bar{R}$. As it contains no hydrogen, it has neither acid nor basic properties, and is therefore without action upon vegetable colors.

63. Water Type.—A molecule of water consists of two atoms of hydrogen linked together by oxygen, thus: H—O—H. By exchanging one of these hydrogen atoms for a negative monad, an acid, $\bar{R}$—O—H, is produced. By a similar exchange for a positive atom, a base, $\overset{+}{R}$—O—H, is obtained. By replacing both the hydrogen atoms, one by a positive, the other by a negative atom, a salt, $\overset{+}{R}-O-\overset{-}{R}$, results. Hence these three classes of bodies are said to be formed upon the plan of structure of water; that is, upon the water type.

Acids and bases may also be viewed as compounds of the monad radical hydroxyl. If hydroxyl $H-O-$ unite with $\overset{-}{R}$, it forms an acid $\overset{-}{R}-O-H$; if with $\overset{+}{R}$, it gives a base $\overset{+}{R}-O-H$. This method of viewing acids and bases is convenient for many purposes.

64. Naming of Acids, Bases, and Salts.—Acids, bases, and salts, like binaries, are named from their constituent atoms. The termination of the negative is changed, however, to indicate that the atoms are linked by oxygen. These negative ternary terminations are universally ate and ite.

EXAMPLES.—Potassium and sulphur when directly united, form potassium sulphide; when united by oxygen, potassium sulphate, sulphite, or hypo-sulphite, according to the equivalence of the sulphur. The binary hydrogen nitride becomes, by the introduction of linking oxygen, hydrogen nitrate or nitrite, both ternary acids. And in the same way, copper hydride becomes copper hydrate.

65. Change in Termination of the Positive Atom.—The positive atom, as in binaries, retains its name unaltered except when it acts with more than one equivalence. It then takes the terminations ous and ic.

EXAMPLES.—Mercurous and mercuric nitrates, cuprous and cupric sulphates; ferrous and ferric phosphates, stannous and stannic silicates, etc.

66. Common Names of Acids and Bases.—As both acids and bases contain oxygen and hydrogen, they are commonly named from the characteristic constituent, giving it the termination ic and ous, according to its equivalence, and adding the word acid or base.

EXAMPLES.—The common name of hydrogen sulphate is sulphuric acid; of hydrogen nitrite, nitrous acid; of hydrogen phosphate, phosphoric acid; of hydrogen hypo-chlorite, hypo-chlorous acid. So too, calcium hydrate is calcic base; zinc hydrate, zincic base; ferric hydrate, ferric base; ferrous hydrate, ferrous base; aluminic hydrate, aluminic base.

67. Formation of Ternary Molecules.—Ternary molecules are formed in two ways:—

1st. By the direct union of binary molecules.

2d. By substitution, from each other.

68. Formation of Ternaries by Direct Union.—Ternaries are formed by the direct union of the oxide of a more positive atom with the oxide of a less positive or negative one. In this case, the name oxide is dropped, and the name of the negative takes ate if it terminated before in ic; or ite if it ended before in ous. Whenever water is the negative oxide, the body produced is a base: when it is the positive, the ternary is an acid.

EXAMPLES.—Sodium oxide and phosphoric oxide unite to form sodium phosphate; here the oxide of both is dropped, and the *ic* of the negative oxide is changed into *ate*. So calcium oxide and sulphurous oxide form calcium sulphite. Silver oxide and hypochlorous oxide form silver hypo-chlorite. Again, when hydrogen oxide—water—unites with sulphuric oxide, hydrogen sulphate, or sulphuric acid, results; when it unites with potassium oxide, potassium hydrate is produced, the water being now negative.

Negative oxides are sometimes called anhydrides, because they may be formed from acids by the abstraction of water.

69. Formation of Ternaries by Substitution.—Ternaries are also formed from each other, by substitution. An acid by exchanging its hydrogen for a positive atom, becomes a salt; a base, exchanging its hydrogen for a more negative substance, becomes also a salt; a salt may become an acid or a base, according as its positive or its negative constituent is replaced by hydrogen; and by exchanging positive for negative atoms or the reverse, bases may be converted into acids or acids into bases.

EXAMPLES.—Hydrogen chlorate or chloric acid, by exchanging its hydrogen for barium, becomes barium chlorate, a salt. Lithium hydrate, or lithic base, by exchanging its hydrogen for carbon becomes lithium carbonate, also a salt. So magnesium silicate becomes hydrogen silicate or silicic acid, by replacing its positive portion by hydrogen, and magnesium hydrate or magnesic base, by replacing its negative portion by the same element.

Whenever a base and an acid are brought in contact, a salt and water are produced. Or, graphically, $\overset{+}{R}-O-H$ and $\overset{-}{R}-O-H$ produce $\overset{+}{R}-O-\overset{-}{R}$ and $H-O-H$.

70. Classification of Acids.—Acids are divided into two classes, called ortho-acids and meta-acids.

1st. Ortho-acids are those acids in which all the oxygen has a linking function. In these acids, therefore, there are as many atoms of hydrogen and of oxygen—i. e., of hydroxyl—as is equal to the equivalence of the negative atom or atomic group.

EXAMPLES.

$Cl^{VII}(OH)_7$ is ortho-perchloric acid.
$Cl^{V}(OH)_5$ is ortho-chloric acid.
$Cl'''(OH)_3$ is ortho-chlorous acid.
$Cl'(OH)$ is ortho-hypo-chlorous acid.

$S^{VI}(OH)_6$ is ortho-sulphuric acid.
$S^{IV}(OH)_4$ is ortho-sulphurous acid.
$S''(OH)_2$ is ortho-hypo-sulphurous acid.

2d. Meta-acids are acids which contain saturating as well as linking oxygen. The oxygen atoms therefore, in a meta-acid, always exceed the hydrogen atoms.

EXAMPLES.

$(ClO)'(OH)$ is meta-chlorous acid or hydrogen meta-chlorite.
$(CO)''(OH)_2$ is meta-carbonic acid or hydrogen meta-carbonate.
$(PO)'''(OH)_3$ is meta-phosphoric acid or hydrogen meta-phosphate.
$(CrO)^{IV}(OH)_4$ is meta-chromic acid or hydrogen meta-chromate.
$(IO)^{V}(OH)_5$ is meta-per-iodic acid or hydrogen meta-per-iodate.

71. Formation of Meta-acids.—Meta-acids are derived from ortho-acids by subtracting from them one or more molecules of water; the acid being mono-, di-, or tri-meta, according to the number of molecules of water taken from the ortho-acid to form it.

EXAMPLES.

$S^{VI}(OH)_6$, less H_2O, leaves $(SO)^{IV}(OH)_4$, mono-meta-sulphuric acid.
$S^{VI}(OH)_6$, less $(H_2O)_2$, leaves $(SO_2)''(OH)_2$, di-meta-sulphuric acid.
$N^{V}(OH)_5$, less H_2O, leaves $(NO)'''(OH)_3$, mono-meta-nitric acid.
$N^{V}(OH)_5$, less $(H_2O)_2$, leaves $(NO_2)'(OH)$, di-meta-nitric acid.
$As'''(OH)_3$, less H_2O, leaves $(AsO)'(OH)$, mono-meta-arsenous acid.

The precise manner in which the molecule is affected by this abstraction of water is represented by the following graphic formulas showing the production of the meta-perchloric acids :—

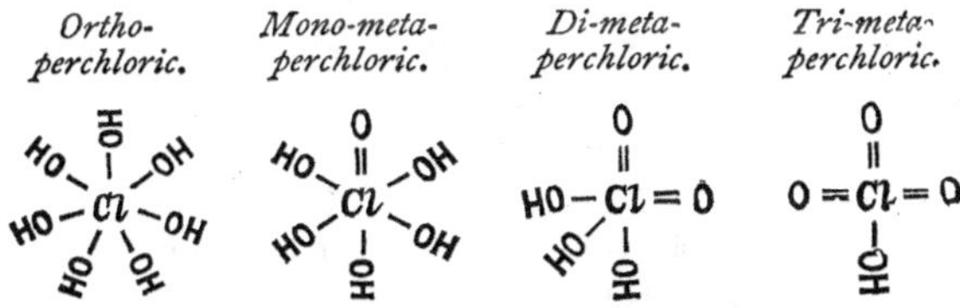

Every molecule of water thus removed, it should be observed, leaves one atom of saturating oxygen. Hence every mono-meta-acid has one such atom, every di-meta-acid two, and every tri-meta-acid three.

TABULAR VIEW OF ORTHO- AND META-ACIDS.

	Monads	*Dyads*	*Triads*	*Tetrads*	*Pentads*	*Hexads*	*Heptads*
Ortho	$H\overset{I}{R}O$	$H_2\overset{II}{R}O_2$	$H_3\overset{III}{R}O_3$	$H_4\overset{IV}{R}O_4$	$H_5\overset{V}{R}O_5$	$H_6\overset{VI}{R}O_6$	$H_7\overset{VII}{R}O_7$
Mono-meta	....	.....	$H\overset{III}{R}O_2$	$H_2\overset{IV}{R}O_3$	$H_3\overset{V}{R}O_4$	$H_4\overset{VI}{R}O_5$	$H_5\overset{VII}{R}O_6$
Di-meta	....	.....	.....	.....	$H\overset{V}{R}O_3$	$H_2\overset{VI}{R}O_4$	$H_3\overset{VII}{R}O_5$
Tri-meta	....	.....	.....	.....	.....	.....	$H\overset{VII}{R}O_4$

72. Basicity of Acids.—The hydrogen in an acid which is linked to the atomic group by oxygen, is called basic hydrogen because it is readily exchanged for a more positive atom or group of atoms. Acids are said to be mono-basic, di-basic, tri-basic, or tetra-basic, according as they contain one, two, three, or four atoms of basic hydrogen. All acids are poly-basic which contain within their molecules more than one of these hydroxyl groups.

EXAMPLES.

$(NO_2)'(OH)$, di-meta-nitric acid, is monobasic.
$(CO)''(OH)_2$, mono-meta-carbonic acid, is dibasic.
$B'''(OH)_3$, ortho-boric acid, is tribasic.

In all ortho-acids the basicity is, of course, equal to the equivalence of the negative atom.

73. Ortho- and Meta-bases.—Like acids, bases may be either ortho- or meta-, and for the same reasons. But, since positive atoms rarely vary in equivalence, but a very few meta-bases are known.

EXAMPLES.

$K'(OH)$ is ortho-potassic base, or ortho-potassium hydrate.
$Ca''(OH)_2$ is ortho-calcic base, or ortho-calcium hydrate.
$Pt^{IV}(OH)_4$ is ortho-platinic base, or ortho-platinic hydrate.
$Fe_2{}^{VI}(OH)_6$ is ortho-ferric base, or ortho-ferric hydrate.
$ZrO(OH)_2$ is meta-zirconic base, or meta-zirconium hydrate.
$Fe_2{}^{VI}O_2(OH)_2$ is di-meta-ferric base, or di-meta-ferric hydrate.

74. Acidity of Bases.—The hydrogen of a base is called acid hydrogen, because replaceable by a negative atom. The acidity of a base depends upon the number of hydroxyl groups, like the basicity of an acid. Bases are mon-acid, di-acid, tri-acid, etc., as acids are monobasic, etc.

EXAMPLES.

Ferrous base $Fe''(OH)_2$ is di-acid.
Argentic base $Ag'(OH)$ is mon-acid.
Aluminic base $Al_2^{VI}(OH)_6$ is hex-acid.

75. Formation of Salts.—Salts are formed from acids by replacing their basic hydrogen by positive atoms.

If the acid has its systematic name, the name of the salt is obtained from it by putting the name of the positive atom in place of the hydrogen.

If the acid has its common name, the name of the salt is formed by placing the name of the positive atom first, followed by that of the acid, the termination ic being changed to ate, and ous to ite.

EXAMPLES.

Hydrogen nitrate, or nitric acid, and sodium, give sodium nitrate.

Hydrogen chlorite, or chlorous acid, and barium, give barium chlorite.

Hydrogen hypo-iodite, or hypo-iodous acid, and zinc, give zinc hypo-iodite.

76. Formulas of Salts.—In writing the formula of a salt, regard must be had, both to the basicity of the acid, and to the equivalence of the replacing atom. As many molecules of the acid must be taken as is necessary to furnish a number of hydrogen-atoms equal to the least common multiple of the basicity of the acid and the equivalence of the replacing atom.

EXAMPLES.—It is required to write the formula of calcium phosphate. Calcium phosphate is derived from hydrogen phosphate—phosphoric acid—by replacing hydrogen by calcium. The formula of mono-meta-phosphoric acid—the most common form, and therefore referred to when no other is specified—is $PO'''(OH)_3$ or more conveniently, H_3PO_4. Calcium is a dyad; the acid is tri-basic; the least common multiple of two and three is six; as many molecules of the acid are required as is necessary to yield six hydrogen atoms; this number is two, written $(H_3PO_4)_2$ or for convenience of replacement $H_6(PO_4)_2$. The six atoms of hydrogen can be exchanged for three of Ca''; making this change, we have $Ca''_3(PO_4)_2$, or calcium phosphate.

Mono-basic Acids.

Na'	and HNO_3	give	$Na'NO_3$	Sodium nitrate.
Ca''	and $(HNO_2)_2$	"	$Ca''(NO_2)_2$	Calcium nitrite.
Bi'''	and $(HClO_3)_3$	"	$Bi'''(ClO_3)_3$	Bismuth chlorate.
Zr^{IV}	and $(HPO_3)_4$	"	$Zr^{IV}(PO_3)_4$	Zirconium phosphate.

Di-basic Acids.

K'_2	and H_2SO_3	give	K'_2SO_3	Potassium sulphite.
Ba''	and H_2CO_3	"	$Ba''CO_3$	Barium carbonate.
Au'''_2	and $(H_2CrO_4)_3$	"	$Au'''_2(CrO_4)_3$	Gold chromate.
Pt^{IV}	and $(H_2SO_4)_2$	"	$Pt^{IV}(SO_4)_2$	Platinic sulphate.

Tri-basic Acids.

Ag'_3	and H_3AsO_4	give	Ag'_3AsO_4	Silver arsenate.
Zn''_3	and $(H_3IO_4)_2$	"	$Zn''_3(IO_4)_2$	Zinc iodate.
Bi'''	and H_3NO_4	"	$Bi'''(NO_4)$	Bismuth nitrate.
Sn^{IV}_3	and $(H_3SbO_3)_4$	"	$Sn^{IV}_3(SbO_3)_4$	Tin antimonite.

Tetra-basic Acids.

Cs'_4	and H_4SiO_4	give	Cs_4SiO_4	Cæsium silicate.
Fe''_2	and H_4SeO_5	"	Fe''_2SeO_5	Ferrous selenate
Au'''_4	and $(H_4CO_4)_3$	"	$Au'''_4(CO_4)_3$	Gold carbonate.
Zr^{IV}	and H_4TeO_4	"	$Zr^{IV}TeO_4$	Zirconium tellurite.

77. Salts Derived from Bases.—Salts may be derived from bases as well as from acids. They are viewed as so derived only in a few cases, where they form a peculiar class of basic salts.

78. Normal, Acid, Basic, and Double Salts.—Normal salts are salts which contain neither basic nor acid hydrogen. They are formed by the complete replacement of the hydrogen of an acid or a base.

Acid salts are those which contain basic

hydrogen. The replacement in the acid being only partial, the salt still acts like an acid, turning vegetable blues to reds.

Basic salts are salts formed by the partial replacement of the hydrogen of a base by a negative atom. They contain therefore acid hydrogen, and often act like a base upon vegetable colors.

Double salts are salts containing two or more different positive atoms.

EXAMPLES.

Normal Salts.	*Acid Salts.*
Potassium chlorate $KClO_3$	Hydro-sodium sulphite $HNaSO_3$
Calcium sulphate $Ca''SO_4$	Hydro-cæsium carbonate $HCsCO_3$
Bismuth phosphate $Bi'''PO_4$	Hydro-barium phosphate $HBaPO_4$
Sodium silicate Na_2SiO_3	Hydro-cupric silicate H_2CuSiO_4

Basic Salts.

Lead hydro-nitrate	$H(NO_2)'PbO_2$
Copper hydro-acetate	$H(Ac)'CuO_2$
Mercuric hydro-iodite	$H(IO)'HgO_2$
Aluminic hydro-silicate	$H_2Si^{IV}Al_2O_6$

Double Salts.

Potassio-sodium selenate	$KNaSeO_4$
Sodio-calcium antimonate	$NaCa''SbO_4$
Baro-zincic silicate	$Ba''Zn''SiO_4$
Cæsio-rubidic carbonate	$Cs'Rb'CO_3$

Mono-basic acids can form only normal salts. Poly-basic acids can form normal, acid, and double salts.

79. Empirical and Rational Formulas.—An empirical or experimental formula is one derived from analysis. It expresses the kind and relative number of atoms in the molecule. It may be also a true molecular formula, in which case it expresses the absolute number of atoms the molecule contains.

A rational formula not only expresses the kind and the absolute number of atoms contained in any molecule, but also indicates how those atoms are arranged. All graphic formulas are rational.

EXAMPLES.—HNO_3, $CaCl_2O_2$, $HgCl$, $CuClO_3$, are all *empirical formulas*, derived from analysis. The first two are molecular, and express the absolute number of atoms; but the *molecular formulas* of the second two are Hg_2Cl_2 and $Cu_2Cl_2O_6$, the molecule being twice as heavy, in each case.

The rational formulas of the four bodies above given, represented graphically, are :—

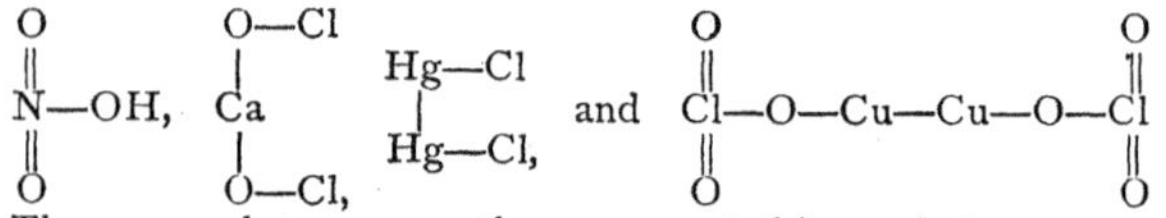

```
O        O—Cl                         O                       O
‖        |        Hg—Cl               ‖                       ‖
N—OH,    Ca       |          and      Cl—O—Cu—Cu—O—Cl
‖        |        Hg—Cl,              ‖                       ‖
O        O—Cl,                        O                       O
```

The same substances are thus represented in symbols :—

$$(NO_2)OH, \quad CaO_2Cl_2, \left\{ \begin{matrix} HgCl \\ HgCl \end{matrix} \right. \text{ and } Cu_2O_2(ClO_2)_2.$$

80. Sulphur and Selenium Acids, Bases, and Salts.—The atoms of ternary molecules may be united by the negative dyads sulphur and selenium, as well as by oxygen. These molecules are named and formulated in precisely the same way as those formed by oxygen, and are distinguished from these by the prefix sulph- or selen-, given to the negative name.

EXAMPLES.

$As(OH)_3$ Hydrogen arsenite	$As(SH)_3$ Hydrogen sulph-arsenite.
$Sb(OAg)_3$ Silver antimonite	$Sb(SAg)_3$ Silver sulph-antimonite.
$SbO(ONa)_3$ Sod. antimonate	$SbS(SNa)_3$ Sod. sulph-antimonate.
$SbO(OK)_3$ Potas. antimonate	$SbSe(SeK)_3$ Potas. selen-antimonate.

The sulphur and selenium in a molecule may be saturating, or linking, or both. These two substances, and also oxygen, may co-exist in the same molecule.

§3. TERNARY MOLECULES UNITED BY TRIADS.

81. Classification of Molecules United by Nitrogen.—The negative triads which may perform a linking function, are nitrogen, phosphorus, and arsenic. Of these, but a few compounds united by phosphorus or arsenic, and these comparatively unimportant, are known. Molecules whose atoms are united by nitrogen are divided into three classes, on the same principle upon which those united by negative dyads are classified. These classes are called amides, amines, and alkalamides.

82. Definition of an Amide.—An amide is made up of molecules consisting of one or more negative atoms united by nitrogen to hydrogen.

The general formula of an amide is $\overset{-}{R}—N=H_2$.

83. Definition of an Amine.—An amine-molecule consists of one or more positive atoms united by nitrogen to hydrogen.

The general formula of an amine is $\overset{+}{R}—N=H_2$.

84. Definition of an Alkalamide.—An alkalamide-molecule contains both positive and negative atoms united by nitrogen.

The general formula of an alkalamide is $\overset{-}{R}—\overset{\overset{\displaystyle H}{|}}{N}—\overset{+}{R}$.

85. Ammonia Type.—An ammonia-molecule consists of three hydrogen-atoms linked together by nitrogen thus, $H—\overset{\overset{\displaystyle H}{|}}{N}—H$. By exchanging a portion of the hydrogen for one or more negative atoms, an amide results; for one or more positive atoms an amine is obtained; and for one positive and one negative, an alkalamide results. These three classes of bodies have a structure similar to that of ammonia. They are said therefore to belong to the ammonia type.

Amides and amines are often regarded as compounds of the monad radical $(H_2N)'$, amidogen. United to $\overset{-}{R}$, it gives an amide; to $\overset{+}{R}$, an amine.

86. Naming of Derived Ammonias. — Amides and amines are called primary, secondary, or tertiary, according as one, two, or three of the hydrogen-atoms are replaced. The individual substances are named by prefixing the Greek numerals to the name of the replacing atom.

EXAMPLES.—Writing ammonia thus, $\left.\begin{matrix}H\\H\\H\end{matrix}\right\}N$, we may have

	Primary.	*Secondary.*	*Tertiary.*
Amide	$\left.\begin{matrix}(CN)'\\H\\H\end{matrix}\right\}N$ *Cyanamide.*	$\left.\begin{matrix}I\\I\\H\end{matrix}\right\}N$ *Din-iodamide.*	$\left.\begin{matrix}Cl\\Cl\\Cl\end{matrix}\right\}N$ *Tri-chloramide.*
Amine	$\left.\begin{matrix}K\\H\\H\end{matrix}\right\}N$ *Potassamine.*	$\left.\begin{matrix}Na\\Na\\H\end{matrix}\right\}N$ *Di-sodamine.*	$\left.\begin{matrix}Rb\\Rb\\Rb\end{matrix}\right\}N$ *Tri-rubidamine.*
Alkalamide	——	$\left.\begin{matrix}K\\Cl\\H\end{matrix}\right\}N$ *Chloro-potassamide.*	$\left.\begin{matrix}K\\K\\Cl\end{matrix}\right\}N$ *Chloro-di-potassamide.*

Amides, amines, and alkalamides, regarded as derivative ammonias, are called mon-amides, di-amides, tri-amides, tetr-amides, etc., according to the number of nitrogen-atoms in the type.

EXAMPLES.

	Mono.	*Di.*	*Tri.*
Amide	$\left.\begin{matrix}(NO_2)'\\H\\H\end{matrix}\right\}N$ *Nitryl-mon-amide.*	$\left.\begin{matrix}(CO)''\\H_2\\H_2\end{matrix}\right\}N_2$ *Carbonyl-di-amide.*	$\left.\begin{matrix}(PO)'''\\H_3\\H_3\end{matrix}\right\}N_3$ *Phosphoryl-tri-amide.*
Amine	$\left.\begin{matrix}Na\\H\\H\end{matrix}\right\}N$ *Sodio-mon-amine.*	$\left.\begin{matrix}Zn''\\H_2\\H_2\end{matrix}\right\}N_2$ *Zinc-di-amine.*	$\left.\begin{matrix}Bi'''\\H_3\\H_3\end{matrix}\right\}N_3$ *Bismuth-tri-amine.*
Alkalàmide	$\left.\begin{matrix}Na\\I\\H\end{matrix}\right\}N$ *Sodio-iod-amide.*	$\left.\begin{matrix}(CO)''\\Hg''\\H_2\end{matrix}\right\}N_2$ *Mercuro-carbonyl-di-amide.*	$\left.\begin{matrix}Bi'''\\(PO)'''\\H_3\end{matrix}\right\}N_3$ *Bismuth-phosphoryl-tri-amide.*

Tertiary derivatives are included here because of their similar origin. Many of them really belong with binary compounds. These bodies are sometimes called *nitriles*, and secondary derivatives are sometimes called *imides*.

87. Mixed Compounds of Hydroxyl and Amidogen.—As those bodies which are formed on the water type may be viewed as binary compounds of hydroxyl, and those which are formed on the ammonia type as similar compounds of amidogen, it is evident that a poly-equivalent radical may combine with both these residues at once, thus forming compounds intermediate between those of the two groups mentioned. If the poly-equivalent radical be negative, the bodies produced are called *amic acids*; if positive, *amid-hydrates*.

EXAMPLES.—Sulphuryl $(SO_2)''$, the radical of sulphuric acid, forms in this way sulphamic acid $(SO_2)''\left\{\begin{matrix}(NH_2)\\(OH)\end{matrix}\right.$. And zinc, a positive dyad, forms $Zn\left\{\begin{matrix}(NH_2)\\(OH)\end{matrix}\right.$ zincamid-hydrate.

TABULAR VIEW OF MOLECULAR STRUCTURE.

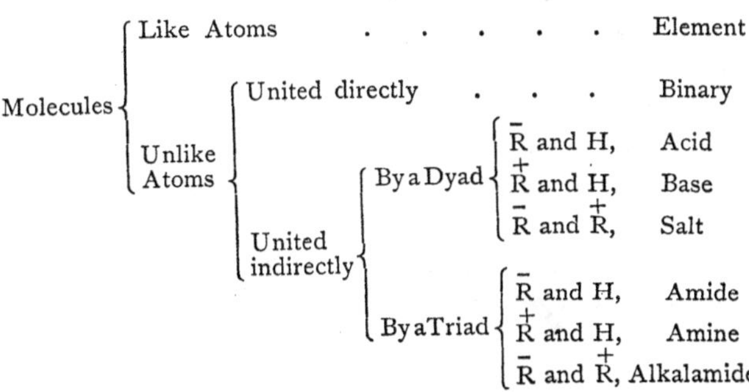

88. Recapitulation. — I. Molecules are of two classes:—

1st. Those composed of like atoms and called Elementary.

2d. Those composed of unlike atoms and called Compound.

II. All compound molecules are of two classes:—

1st. Those whose atoms are directly united, called Binary.

2d. Those whose atoms are indirectly united, called Ternary.

III. Ternary molecules are of two classes:—

1st. Those in which the linking atom is a dyad, called Hydrates.

2d. Those in which the linking atom is a triad, called Compound Ammonias.

IV. Hydrates are divided into three classes:—

1st. Acid hydrates, or acids, which consist of a negative atom, or radical, united to hydrogen.

2d. Basic hydrates, or bases, which consist of a positive atom, or radical, united to hydrogen.

3d. Salts, which consist of a positive atom, or radical, united to a negative atom or radical.

V. Compound Ammonias are divided into three classes:—

1st. Amides, containing a negative radical united to hydrogen.

2d. Amines, containing a positive radical united to hydrogen.

3d. Alkalamides, containing a positive radical united to a negative radical.

EXERCISES.

§ 1.

1. How many atoms may a compound molecule contain?
2. How is the molecular weight of such a molecule obtained?
3. How many bonds may there be in a compound molecule? How many perissad atoms?
4. Define a Binary molecule. A Ternary molecule.
5. Give the rule for naming Binaries. Illustrate it.
6. Platinum forms two compounds with bromine; name them.
7. What atoms have more than two equivalences?
8. What are the oxides of phosphorus? The chlorides?
9. What are formulas, and how are they written?
10. Give the formula of potassium iodide; of lead sulphide; of phosphorus nitride; of calcium chloride; of gold oxide; of silver arsenide; of silicic bromide; of antimonous oxide.
11. Give the names of NaCl, SrO, BiP, Cu_3As_2, $(CS_2)_2$, $(SnO)_4$.
12. How do atoms of different equivalences combine?
13. How are compound molecules formed from elemental ones?
14. Define compound radicals. How are they named?
15. How is variation in atomic equivalence explained?

§ 2.

16. How are ternary molecules united? By what dyads?
17. Define, and explain the general formula of, an acid, a base, and a salt.
18. What is the water type?
19. What compounds are formed by the union of potassium to monad, triad, pentad, and heptad chlorine successively, by oxygen?
20. What are the constituents of silver phosphate; lithium carbonate; zinc hydrate; hydrogen bromate; manganous phosphate; mercurous nitrate?
21. What are the formulas of silicic, chromic, iodous, carbonic, hypo-sulphurous, bromous, and titanic acids?
22. Illustrate the formation of ternaries by direct union.

23. What is produced when lead oxide and nitrous oxide unite? bismuthous and selenic oxides? hydrogen and sulphuric oxides?

24. How is barium sulphite produced by substitution? Mercurous hypo-chlorite?

25. Define ortho- and meta-acids. How are the latter derived?

26. Which is H_3PO_4? $HClO$? $HAsO_2$? H_2SeO_2? H_4WO_5?

27. Give the general formula of the mono-meta-acid of a triad; of a pentad; of a hexad.

28. Define basicity of acids. What is a tetra-basic and a poly-basic acid?

29. What is a poly-acid base? What is $HNaO$? H_2BaO_2? $HAuO_2$? H_3BiO_3?

30. How is the name of a salt derived from that of an acid?

31. Give the rule for writing salt-formulas.

32. Write the formulas of sodium bromate; calcium hypo-chlorite; platinic antimonate; stannic chromate; potassium borate; lead arsenite; manganous carbonate.

33. Define and illustrate normal, acid, basic, and double salts.

34. Distinguish an experimental from a rational formula.

35. Give the empirical, rational, and graphic formulas of calcium nitrate. Of mercurous phosphate.

36. Give the formula of silver sulpho-stannate.

§ 3.

37. What are amides? Amines? Alkalamides? How are they named? To what type do they belong?

38. What is a primary amine? A tertiary di-amide?

39. Write the formula of sodamine, chloramide, sulphamide, phosphamic acid.

CHAPTER FOURTH.

VOLUME-RELATIONS OF MOLECULES.

§ 1. Relation of Density to Atomic Weight.

89. Molecular Volume.—By the law of Ampère, all molecules have the same size; that is, all molecules when in the gaseous state occupy the same volume. Every molecular formula therefore, not only expresses the weight of the molecule, but also the volume which it occupies. The molecule of hydrogen is taken as the standard of molecular volume; but, as it is sometimes convenient to speak of atomic volume—which, since no atom can exist free, must be a fiction—the volume occupied by the atom of hydrogen is taken as unity; the volume occupied by the molecule will therefore, be two. Moreover, as all molecules occupy the same volume, the molecular volume of all bodies is assumed to be two, also.

90. Relation of Molecular Weight to Density.—Since molecular weight represents the weight of two volumes, and density represents the weight of one, the density of any homogeneous substance in the state of gas, is one half its molecular weight.

Examples.—Ammonia gas, whose molecular formula is H_3N, has a molecular weight of $3 + 14$ or 17. By the above rule, its *calculated density* is $17 \div 2$ or 8·5; i. e., it is 8·5 times heavier than hydrogen. An experiment with the balance shows that one liter of ammonia gas weighs 0·7627 grams. As one liter of hydrogen gas weighs 0·0896 grams, the *experimental density* of ammonia gas is 0·7627 ÷ 0·0896 or 8·5.

91. Molecular Weight Fixed by Density.—Conversely, knowing the density, the molecular weight may be obtained by doubling it.

The analysis of any homogeneous substance gives the ratio of the constituents only, not their absolute weight. On analytical grounds, therefore, several molecular weights, all multiples of the lowest, may be attributed to the body analyzed. By taking the density of the substance in the state of vapor, however, and doubling it, the true molecular weight is determined.

EXAMPLES.—The analysis of hydrogen oxide—water—shows that in 100 parts of it there are 88·89 parts of oxygen and 11·11 parts of hydrogen. The ratio of 11·11 : 88·89 is 1 : 8. If the molecule contain one part by weight of hydrogen and eight parts by weight of oxygen, its molecular weight will be 1+8 or 9. But it may contain two, three, four or five times this quantity of each constituent, and yet yield the same analytical results. Its molecular weight, so far as analysis goes, may be 9, 18, 27, 36, 45, etc. Upon weighing now, a liter of water-gas—steam—it is found to weigh 0·8047 grams; whence its density is 0·8047 ÷ 0·0896, or 9, and its molecular weight is 9×2 or 18. Water therefore consists of 2 parts by weight of hydrogen and 16 parts by weight of oxygen in each molecule.

Again, olefiant gas—a compound of carbon and hydrogen—affords on analysis 85·71 per cent of carbon and 14·29 per cent of hydrogen, which is the ratio of 6 : 1. There may be then 6 parts by weight of carbon to one of hydrogen in the molecule, in which case the molecular weight of olefiant gas will be 7; or 12 parts of carbon to two of hydrogen, when it will be 14; or 18 to 3, giving 21; or 24 to 4, giving 28; the ratio in all these cases being the same. By experiment, one liter of olefiant gas weighs 1·252 grams. Its density therefore, is 1·252 ÷ 0.0896, or 14, and its molecular weight 14×2 or 28. Hence one molecule of olefiant gas contains 24 parts by weight of carbon and 4 parts by weight of hydrogen.

92. Aid in Determining Atomic Weight.—The atomic

weight of any simple substance is the smallest quantity of it by weight which can enter into the formation of a molecule. By ascertaining therefore, the molecular weights of various compounds of the same element, and by comparing together the quantities of this element by weight which they severally contain, it is easy to fix its atomic weight.

EXAMPLES.—A molecule of water contains 16 parts by weight of oxygen; a molecule of carbonic gas, 32 parts; a molecule of sulphuric oxide, 48 parts; etc. In no known compound, however, is there less than 16 parts by weight of oxygen in a molecule; 16 is therefore the atomic weight of oxygen.

Again, a molecule of marsh-gas contains 12 parts by weight of carbon, a molecule of olefiant gas 24, a molecule of glycerin 36, a molecule of tartaric acid 48, a molecule of citric acid 72. The smallest of these numbers, 12, is therefore the atomic weight of carbon; and the bodies above mentioned contain in each molecule, one, two, three, four, five, and six atoms of carbon, respectively.

§ 2. RELATION OF GASEOUS DIFFUSION TO ATOMIC WEIGHT.

93. Gaseous Diffusion.—In 1825, Döbereiner, having collected some hydrogen gas in a cracked jar, standing over water, noticed that the level of the water within the jar rose one and a half inches in twelve hours; a result obviously due to the escape of the hydrogen through the crack. Graham found, on repeating the experiment, that as the hydrogen escaped outward, a portion of air, much less in amount, entered the jar. And on investigation he ascertained that gases, even when separated by porous partitions, pass freely into each other. This mutual passage of one gas into another is called diffusion.

94. Graham's Law of Diffusion.—Graham showed experimentally that the rapidity with which different gases diffuse into each other, varies for each gas ; and also that this rapidity stands in intimate relation to the density of the gas. This relation is expressed in the following law :—

The velocity of the diffusion of any gas is inversely proportional to the square root of its density.

95. Explanation of Diffusion.—Physics assumes that all gaseous molecules are in rapid motion in straight lines. Now, as the pressure upon all gaseous volumes is equal—being the atmospheric pressure—and as this pressure is exactly balanced by the elasticity of the gas —due to the direct outward impact of its molecules—it follows that the impact of all gaseous molecules is equal.

Moreover, in mechanics, impact is proportional to the product of the mass into the square of the velocity. Since, therefore, molecules differ in weight, it is obvious that lighter ones must move faster than heavier ones, to produce the same effect. If one molecule is four times as heavy as another, the latter needs to move twice as fast to strike the same effective blow; since 4 times 1^2 is the same as once 2^2, both being 4. By the supposition, the weights of the molecules are as 4 : 1 ; their velocities therefore must be as 1 : 2. Hence, their velocities are inversely as the square roots of their molecular weights ; or, what is the same thing, of their densities ; which is the law of diffusion, deduced by Graham from experiment.

EXAMPLES.—The density of oxygen being 16, a molecule of oxygen is 16 times as heavy as a molecule of hydrogen. The elasticity of both gases under the atmospheric pressure, is the same. But, that the impact should be the same, the lighter or hydrogen molecule, must move 4 times as fast as the heavier, or oxygen molecule, since

OXYGEN.			HYDROGEN.	
Wt.	*Veloc.*		*Wt.*	*Veloc.*
$16 \times$	1^2	$=$	$1 \times$	4^2.

That is, hydrogen molecules should move with four times the velocity of oxygen molecules.

Using a thin graphite plate as a partition, Graham found experimentally, that, taking air as unity, the ratio of the diffusion of oxygen is to that of hydrogen as 0·95 : 3·83. But 0·95 : 3·83 : : 1 : 4. As a matter of fact, therefore, hydrogen molecules do move 4 times faster than oxygen molecules.

96. Determination of Molecular Weight by Diffusion.—If the velocity of diffusion of any gas is equal to the inverse square root of its density, then the density of any gas must be equal to the inverse square of its velocity of diffusion. Or mathematically,—calling V the velocity of diffusion and D the density—if $V = \frac{1}{\sqrt{D}}$, then $D = \frac{1}{V^2}$. Of course, by doubling the density thus given, the molecular weight is obtained.

EXAMPLES.—By Graham's table of diffusibilities, carbonic gas has a diffusive power of 0·812 as compared with air, or of 0·212 as compared with hydrogen. But $D = \frac{1}{V^2}$; hence $D = \frac{1}{(\cdot 212)^2} = 22\cdot 22$, The density of carbonic gas being 22·22, its molecular weight must be $22\cdot 22 \times 2$ or 44·44. Analysis shows it to be 22, 44, 66, 88, etc. Diffusion fixes it as the second of these numbers. It was in this way that Soret determined the density, and thence the molecular weight, of ozone.

§ 3. COMBINATION BY VOLUME.

97. Law of Combination by Volume.—The proportions in which gaseous volumes enter into combination, were first investigated by Gay-Lussac. His law asserts:—

1st. That the ratio in which gases combine by volume is always a simple one; and 2d. That the volume of the resulting gaseous product bears a simple ratio to the volumes of its constituents.

98. Deduction of this Law.—The law of combination by volume, which, in Gay-Lussac's time, was purely experimental, has been recently shown by Clausius to be a very simple deduction from the law of Ampère.

According to Ampère's law, equal volumes of all gases contain the same number of molecules. If, therefore, the number of molecules be in any way diminished, the volume itself will be diminished proportionally. Suppose now, that in a given volume of any gas, each molecule is di-atomic, i. e., contains but two atoms; then, if by any means the molecule can be made tetr-atomic, i. e., four-atomed,—the absolute number of atoms remaining the same—the number of molecules will be reduced one half, since each molecule contains twice as many atoms. But by this reduction in the number of molecules a corresponding diminution in volume takes place, and the volume of the gas is reduced one half also.

Again, if the di-atomic molecule became tri-atomic, the number of molecules would be reduced by one third. Hence, the volume originally occupied by these molecules would be reduced in the same ratio.

99. Application of Clausius's Theory.—To apply this reasoning to the facts of volume-combinations, let us consider separately, the combinations which hydrogen forms with the four equivalence-groups, monads, dyads, triads, and tetrads, supposing all their molecules to be di-atomic.

First case.—In the case of monads, one atom combines with one atom of hydrogen ; and since the molecules of both are di-atomic, a molecule will combine with one molecule, and a volume with one volume of hydrogen. All bodies consisting of di-atomic molecules made up of monad atoms, combine, therefore, in equal volumes.

Further, when the monad atom and the hydrogen-atom combine, they form a di-atomic molecule precisely like a molecule of either of its constituents, except that its atoms are unlike. The two di-atomic simple molecules form two di-atomic compound molecules. Two volumes of simple gases give two volumes of a compound gas.

Substances of the first class, then,—i. e., monads—combine with each other volume to volume, and yield two volumes of the product.

Examples.—A chlorine atom Cl, unites with a single hydrogen atom H, to form HCl, both being monads. As the molecules of both are di-atomic, these substances unite molecule to molecule, or volume to volume. On mixing the volume of chlorine with the volume of hydrogen and exposing them to sunlight, they unite to form hydrogen chloride gas, each molecule of which is di-atomic, containing one chlorine atom and one hydrogen atom. The number of molecules after the union being the same as before, the volumes are unaffected. To represent it molecularly :—

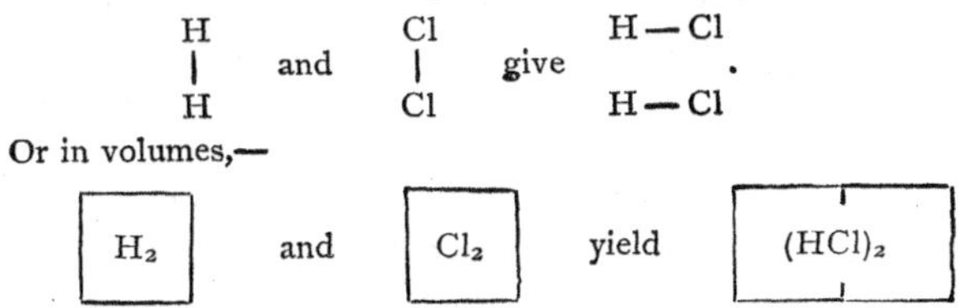

Or in volumes,—

Second case.—If the atom taken be a dyad, then it will unite with two atoms of hydrogen ; or one molecule will unite with two molecules, or one volume with two volumes. Dyads therefore, combine with monads, in the ratio of one volume to two volumes.

Moreover, the molecule which results from the union of one dyad atom with two monad atoms will be tri-atomic. As before union they were di-atomic, three molecules then, make but two now. The total number of molecules is one-third less than before; and, of course, the volume is diminished in the same ratio. Two volumes of one gas and one volume of the other give three volumes; after combination, but two volumes remain. So that three volumes of simple gases give two volumes of a compound gas, a condensation of three volumes to two taking place during union.

Substances of the second class, then,—i. e., dyads—combine with monads in the ratio of one volume to two, and yield two volumes of the product.

EXAMPLES.—The atom of oxygen is bivalent; its molecule is di-atomic. Assuming a fixed number of molecules in the given volume, say 100, then the 200 atoms in these 100 molecules, will unite with 400 atoms, or 200 molecules of hydrogen, producing 600 atoms. In other words, one volume of oxygen will combine with two volumes of hydrogen. Now the water-molecule which results, contains 3 atoms, 2 of them hydrogen and one oxygen; the 600 atoms of these substances above mentioned, will therefore give 200 water molecules, which, by the assumption above, occupy two volumes. Hence one volume of oxygen and two volumes of hydrogen yield two volumes of water-gas.

THIRD CASE.—One triad atom unites with three monad atoms, one molecule with three molecules, one volume with three volumes. The original simple molecules contain two atoms, the resulting compound molecule, four; the number of molecules, and hence the corresponding volume, is therefore reduced one-half, four volumes being condensed into two.

Substances of the third class, i. e., triads, unite with monads in the ratio of one volume to three, and yield two volumes of the product.

Examples.—One atom of nitrogen unites with three atoms of hydrogen to form ammonia. One molecule of nitrogen and three molecules of hydrogen give two molecules of ammonia.

$$\begin{array}{c} N \\ ||| \\ N \end{array} \text{ and } \begin{array}{ccc} H & H & H \\ | & | & | \\ H & H & H \end{array} \text{ give } \underbrace{H\ H\ H}_{N} \quad \overbrace{H\ H\ H}^{N}$$

Four di-atomic give two tetr-atomic molecules. Hence, one volume of nitrogen and three volumes of hydrogen form two volumes of ammonia gas.

Fourth case.—Lastly, one tetrad atom unites with four monad atoms, one tetrad molecule with four monad molecules, one volume of any tetrad with four volumes of any monad. The resulting molecule contains five atoms, and hence, the five original volumes are condensed to two.

Substances of the fourth class, i. e., tetrads, unite with monads in the ratio of one volume to four, and yield two volumes of the product.

Examples.—One atom of carbon and four atoms of hydrogen unite to form marsh gas. That is, C_2 and $(H_2)_4$ give $(H_4C)_2$; or five di-atomic give two pent-atomic molecules. Hence one volume of carbon gas and four volumes of hydrogen form two volumes of marsh gas.

100. Recapitulation of Volume-combinations.—The various ratios in which combinations by volume take place according to Gay-Lussac's law, may be thus represented:—

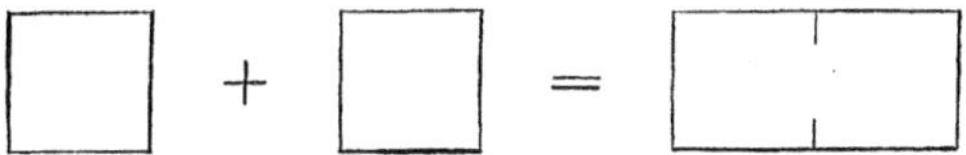

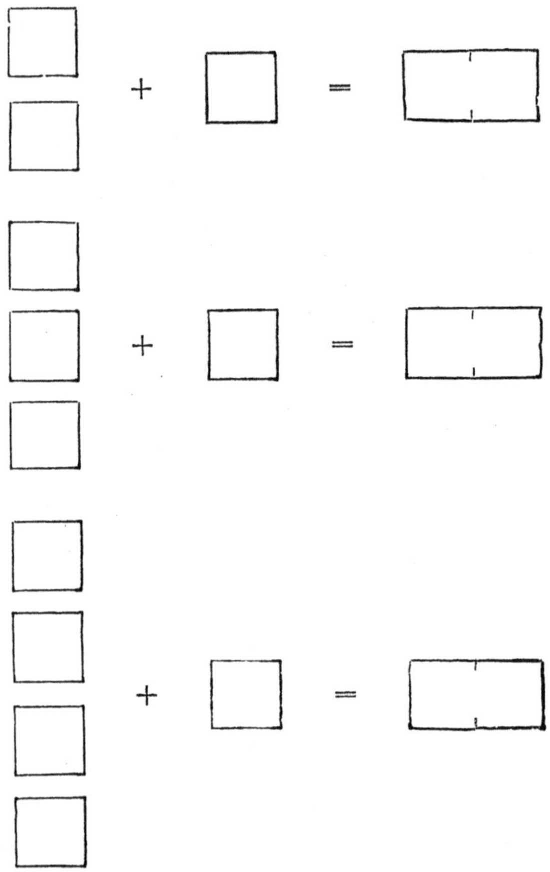

These results correspond precisely with those required by the law of combination by equivalence, which, for the union of di-atomic molecules, gives the following four cases :—

1 molecule and 1 molecule give 2 molecules.
1 molecule and 2 molecules give 2 molecules.
1 molecule and 3 molecules give 2 molecules.
1 molecule and 4 molecules give 2 molecules.

Or, written out fully, to express the atomic character of the molecule :—

Molecules Di-atomic.

1 monad molecule and 1 monad molecule give 2 di-atomic molecules.
1 dyad molecule and 2 monad molecules give 2 tri-atomic molecules.
1 triad molecule and 3 monad molecules give 2 tetr-atomic molecules.
1 tetrad molecule and 4 monad molecules give 2 pent-atomic molecules

101. Cases where Molecules are not Di-atomic.—These cases are practically but two in number; one, where the molecule is mon-atomic, the other, where it is tetr-atomic.

FIRST CASE.—All known mon-atomic molecules are dyads. Hence the atomic combination with monad atoms would be:—

1 atom and 2 atoms give 3 atoms.

And by molecules :—

1 molecule and 1 molecule give 1 molecule.

Or by volumes :—

1 volume and 1 volume give 1 volume.

That is, mon-atomic dyad molecules combine with di-atomic monad molecules in the ratio of equal volumes, yielding one volume of the product.

EXAMPLES.—The dyad, zinc, has a mon-atomic molecule. One atom of zinc unites with two atoms of chlorine to form zinc chloride; that is, one mon-atomic zinc molecule unites with one di-atomic molecule of chlorine, to form one tri-atomic molecule of zinc chloride. As all molecules have the same size, this is equivalent to saying that one volume of zinc-vapor and one volume of chlorine gas combine to give one volume of zinc chloride vapor, a condensation of one-half.

SECOND CASE.—All known tetr-atomic molecules are triads. The atomic combination with monads would therefore be:—

1 atom and 3 atoms give 4 atoms.

And by molecules:—

1 tetr-atomic molecule and 6 di-atomic molecules give 4 tetr-atomic molecules.

Or by volumes:—

1 volume and 6 volumes yield 4 volumes.

That is, tetr-atomic triad molecules combine with di-atomic monad molecules in the ratio of one volume to six volumes, yielding four volumes of the product, a condensation of seven volumes to four.

EXAMPLES.—Phosphorus is a triad, having a tetr-atomic molecule. When it unites with chlorine, we have atomically, P and Cl_3 give PCl_3; molecularly, P_4 and $(Cl_2)_6$ give $(PCl_3)_4$, or by volume:—

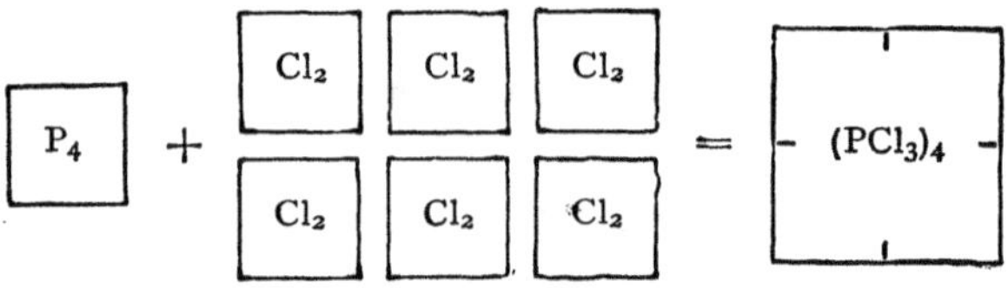

$$\boxed{P_4} + \boxed{Cl_2}\boxed{Cl_2}\boxed{Cl_2}\boxed{Cl_2}\boxed{Cl_2}\boxed{Cl_2} = \boxed{(PCl_3)_4}$$

Or, as sometimes given:—

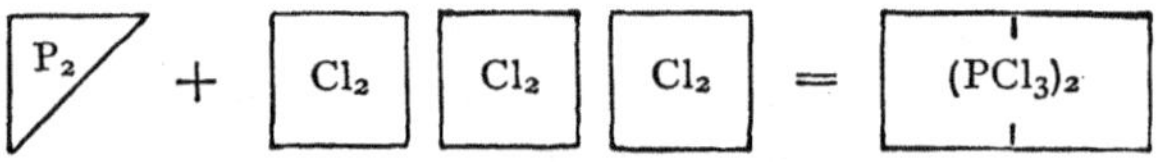

$$P_2 + \boxed{Cl_2}\boxed{Cl_2}\boxed{Cl_2} = \boxed{(PCl_3)_2}$$

102. Tri-atomic and Hex-atomic Molecules.—The law of combination by volume for tri-atomic and hex-atomic molecules, should any substances be found to combine in this way, can easily be deduced from the principles already given.

The entire foregoing chapter furnishes an excellent illustration of the intimate mutual relations between

Physics and Chemistry. Assuming either the physical or the chemical data at pleasure, the other can be deduced from it.

EXERCISES.

§ 1.

1. What is molecular volume, and how is it expressed?
2. One liter of carbonic gas (CO_2) weighs 1·97 grams; what is its calculated, and what its experimental, density?
3. Of what assistance is density in fixing a molecular weight?
4. Iron chloride contains 34·46 per cent of iron, and 65·54 per cent of chlorine; its density is 162·5; what is its molecular weight? How much iron and how much chlorine is there in each molecule?
5. How is an atomic weight determined by density?
6. Hydrogen bromide contains 1·24 per cent of hydrogen, and 98·76 of bromine; its density is 40·5. Mercury bromide contains 55·56 per cent of mercury and 44·44 of bromine; its density is 180. Boron bromide contains 4·38 per cent of boron, 95·62 of bromine; its density is 125·5. Silicon bromide contains 8·04 per cent of silicon and 91·96 of bromine; its density is 174. What is the atomic weight of bromine?

§ 2.

7. What is gaseous diffusion? Give Graham's law.
8. How is the fact of diffusion explained? Illustrate.
9. How is molecular weight fixed by diffusion?

10. Marsh-gas has a diffusibility of 0·35, that of hydrogen being 1, what is its molecular weight?

§ 3.

11. Give Gay-Lussac's law of combination by volume.

12. How may this be derived from the law of Ampère?

13. How do monads combine with hydrogen by volume? Dyads? Triads? Tetrads?

14. In what proportion by volume, do tetrad sulphur and oxygen unite? What volume has the product?

15. Do the volume-combinations as deduced from the law of equivalence, agree with those observed by Gay Lussac?

16. How do mercury and arsenic unite by volume?

CHAPTER FIFTH.

CHEMICAL REACTIONS. STOICHIOMETRY.

§ 1. Chemical Equations.

103. Molecular Stability.—All material molecules are more or less liable to chemical change. The atoms within them may be altered in kind, in number or in relative position, by various external influences. A molecule is the more stable in proportion as it resists this tendency to change.

104. Chemical Reactions.—Any change which takes place in the atoms composing a molecule is called a Chemical Reaction. Both of the substances acting are called Re-agents.

Examples.—When a candle burns, the wax and the oxygen of the air act mutually upon each other, yielding gaseous products entirely unlike the wax or the oxygen. When gunpowder explodes the various molecules which it contains *react* upon each other, and a new set of products is the result. When the components of a Seidlitz powder are mixed together and moistened, they *react* upon each other, producing the well-known effervescence.

105. Reactions Always Molecular.—Chemical reactions always take place within the molecule. When therefore, two substances react upon each other, the changes which result may be viewed as taking place between single molecules. Moreover, since all molecules in homogeneous matter are alike, and what is true of one molecule, is true of any mass of them,—it follows that a molecular change represents accurately, a mass-change.

106. Reactions Expressed by Formulas.—Every formula in Chemistry represents a molecule; and as all reactions are viewed as taking place between molecules, these reactions may be represented by the use of molecular formulas.

107. Chemical Equations.—Chemical reactions are usually represented in the form of equations; the substances entering into the reaction—called factors—constituting the first member, and those issuing from it—called products—the second.

When two molecules act upon each other, the equation representing the reaction may be written by the following rule:—

Place the formulas of the factors—connected by the sign plus—as the first member of the equation, and the formulas of the products—also connected by the sign plus—as the second.

EXAMPLES.—The reaction of the two molecules AB and CD would be represented thus:—

$$AB + CD = AD + CB.$$

Sometimes, though rarely, the minus sign is used in an equation, thus:—

$$ABC - B = AC.$$

108. Weight of the Factors and Products Equal.—As each formula represents a definite weight of matter—the molecular weight—it follows that the quantities of matter taking part in a chemical change are definite in weight. And, moreover, since the atoms are the same after the reaction as before it—being only differently associated—it also follows that no loss of weight can be the result of any chemical reaction.

The sum of the molecular weights of the products must, therefore, always equal the sum of the molecular weights of the factors.

109. Meaning of the Signs.—The equality sign indicates the equality in weight of both members of the equation. The plus sign means simply "and," and signifies that the molecules thus united are mixed together. The minus sign means "from," and indicates the removal of a simpler group of atoms from a more complex one.

110. Classification of Reactions.—Chemical reactions are usually divided into three classes, as follows :—

1st. Analytical reactions; which represent the separation of a complex molecule into simpler ones.

2d. Synthetical reactions; which represent the union of two or more simple molecules, to form a more complex one.

3d. Metathetical reactions; which represent a transposition or exchange of atoms between molecules.

EXAMPLES.—*Analytical reactions* may be represented by the general equation :—

$$AB = A + B.$$

Or, to take an actual example,

$(NH_4)NO_3$	=	N_2O	+	$(H_2O)_2$
Ammonium nitrate.		*Nitrogen oxide.*		*Water.*

which is read thus :—One molecule of ammonium nitrate yields one molecule of nitrogen oxide and two molecules of water.

Synthetical reactions are the reverse of analytical; they are represented by the general equation :—

$$A + B = AB.$$

Or, as an actual example,

CaO	+	SO_3	=	$CaSO_4$
Calcium oxide.		*Sulphuric oxide.*		*Calcium sulphate.*

read thus:—One molecule of calcium oxide and one molecule of sulphuric oxide yield one molecule of calcium sulphate.

Metathetical reactions—from μετατίθημι, to displace or transpose —are represented by the general formula:—

$$\overset{+}{A}\overset{-}{B} + \overset{+}{C}\overset{-}{D} = \overset{+}{A}\overset{-}{D} + \overset{+}{C}\overset{-}{B}.$$

Or, practically, by the equation:—

$$\underset{\textit{Silver nitrate.}}{Ag(NO_3)} + \underset{\textit{Sodium chloride.}}{NaCl} = \underset{\textit{Silver chloride.}}{AgCl} + \underset{\textit{Sodium nitrate.}}{Na(NO_3)}$$

A molecule of silver nitrate reacts upon a molecule of sodium chloride, to produce one molecule of silver chloride and one of sodium nitrate; an exchange taking place between positive atoms.

In all the above examples, each letter in the general equations, and each formula in the special, represents an entire molecule. Less than an entire molecule cannot enter into, or issue from, any chemical reaction.

111. Conditions Favoring Chemical Change.—Facility of chemical change depends, to a large extent, upon the ease with which the atoms of any molecule may be re-arranged. It is found, for example, that chemical changes take place very readily when the substances acting are in the liquid or gaseous state. Hence fusion, or solution, by which bodies are liquefied, or vaporization, by which they are converted into gases, facilitate chemical action.

112. Berthollet's Laws.—Those conditions of chemical change which depend upon solubility, are stated in the following general law, first established by Berthollet:—

Whenever, on mixing two substances in solution, a compound can be formed by a re-arrangement of their atoms, which is insoluble in the menstruum employed, such compound will be formed, and will appear as a precipitate.

EXAMPLES.—If AB be dissolved in water, and CD, also dissolved in water, be added to it, then any re-arrangement must obviously produce AD and CB. AB and CD are soluble in water; but AD, or CB, or both, may be insoluble. In either case, the new and insoluble compound will separate from the solution in the solid form, the liquid losing its clearness, and becoming turbid.

The solid substance which thus separates from a solution, is called a precipitate. Any substance which will produce a precipitate when added to a solution of any other substance, is called a precipitant. The process of producing a precipitate is called precipitation.

113. Precipitation both Chemical and Physical.—The first step in precipitation is the re-arrangement of the atoms; this is a chemical result. The second step is the separation of the insoluble product; this depends on the adhesion between the liquid molecules and those of the solid, and hence is a physical result. No conclusions can be drawn, therefore, as to the strength of the chemism, from the mere facts of precipitation.

114. Law of Gaseous Change.—The second law of Berthollet holds when the product of the reaction, instead of being a solid and insoluble, is a gas. It may thus be stated:—

Whenever, by the action of bodies upon each other, any substance, volatile at the temperature of the experiment can be formed by a re-arrangement of the atoms, such re-arrangement will take place, and such substance will be evolved as a gas or vapor.

EXAMPLES.—Theoretically, as above, AB and CD, by re-arrangement, give AD and CB. If AD or CB is volatile at the temperature of the experiment, it will separate from the solution in the gaseous form. Or, actually, let $Na_2(CO_3)$ and $H_2(SO_4)$ be mixed together in solution. By the chemical re-arrangement, $Na_2(SO_4)$

and $H_2(CO_3)$ will be produced. As, however, the body $H_2(CO_3)$ cannot exist at ordinary temperatures, it separates into H_2O and CO_2; which latter substance, being a gas, escapes from the solution.

Again, on mixing together potassium nitrate $K(NO_3)$ and hydrogen sulphate $H_2(SO_4)$, there will result by the re-arrangement, the bodies $HK(SO_4)$ and $H(NO_3)$—hydro-potassium sulphate, and hydrogen nitrate; this being the chemical part of the change. If now heat be applied to the mixture, the nitric acid, being volatile, will escape as a vapor.

The rapid escape of a gas from a liquid, such as is noticed in mixing Seidlitz powders, for example, is called effervescence.

115. Prediction of Results.—Whether a chemical change will actually take place or not, may, in many cases, be predicted by means of these laws, if the properties of the products be known. By the use of a table—given in the appendix—showing the solubility of various substances, all cases under the first law may be predicted; and by having some familiarity with the volatility of various bodies, cases may be predicted under the second law.

EXAMPLES.—If calcium chloride and sodium carbonate be mixed in solution, will there be a precipitate? The reaction is thus written:—

$$CaCl_2 + Na_2(CO_3) = Ca(CO_3) + (NaCl)_2$$

Referring to the table, it will be seen that $Ca(CO_3)$ calcium carbonate, is insoluble. It will therefore separate in the solid form and fall as a precipitate.

If however, an acid be at the same time formed, and the solid substance be soluble in acids, there will be no precipitate. If hydrogen sulphide and ferrous sulphate be mixed in solution, the reaction would be represented by the equation:-

$$H_2S + Fe(SO_4) = FeS + H_2(SO_4)$$

By the table, ferrous sulphide (FeS), though insoluble in water, is soluble in acids. Were water alone present, this substance would be precipitated; but as sulphuric acid ($H_2(SO_4)$) is in the solution, being set free in the reaction, there will be no precipitate.

Again, if ammonium sulphate and calcium carbonate be heated together, will there be a change? The only possible exchange is the following:—

$$(NH_4)_2(SO_4) + Ca(CO_3) = (NH_4)_2(CO_3) + Ca(SO_4)$$

Ammonium sulphate. Calcium carbonate. Ammonium carbonate. Calcium sulphate.

But ammonium carbonate is volatile; it will therefore escape in vapor, leaving the calcium sulphate behind.

116. Modes of Chemical Action.—Chemical changes in matter may take place in five different ways, namely:—

1. By the direct union of simpler molecules to form a more complex one.
2. By the separation of a complex molecule into simpler ones.
3. By the substitution in a molecule of one atom or group of atoms for another or for several others.
4. By the mutual exchange of atoms between molecules.
5. By the re-arrangement of the atoms within a single molecule.

EXAMPLES.—1st. All synthetical reactions belong to the first class of chemical changes; as:—

$$Zn + Cl_2 = ZnCl_2$$

Zinc. Chlorine. Zinc chloride.

2d. All analytical reactions belong to the second class; as:—

$$H_2SO_4 = H_2O + SO_3$$

Hydrogen sulphate. Water. Sulphuric oxide.

3d. Substitution reactions; as:—

$$C_2H_6 + Cl_2 = C_2H_5Cl + HCl$$

Ethyl hydride. Chlorine. Ethyl chloride. Hydrogen chloride.

4th. All metathetical reactions represent the fourth class; as:—

$$CaCl_2 + K_2(C_2O_4) = Ca(C_2O_4) + (KCl)_2$$

Calcium chloride. Potassium oxalate. Calcium oxalate. Potassium chloride.

5th. The conversion of ammonium cyanate into urea :—

$$(NH_4)(CNO) = (CO)(H_2N)_2$$

Ammonium Cyanate. *Urea.*

117. Energy of Chemical Attraction.—The energy with which chemical action takes place when the proper conditions are secured, is due to the strength of the attraction between the atoms. Within its own limits, this atomic attraction is irresistible.

EXAMPLES.—Oxygen and hydrogen gases have never been condensed to liquids by the greatest mechanical pressure obtainable; yet under the influence of chemical attraction, they unite to form a perfectly stable liquid.

Again, the awful energy developed when gunpowder or nitro-glycerin explodes, is due simply to the strength of the attraction which re-arranges their atoms.

118. Connection of Physical Force with Chemical Action.—Physical forces are molecular motions. Under given circumstances they all may bring about chemical action. So, on the other hand, every chemical action produces a physical result, evolving heat, light or electricity. In other words, some mode of force is necessary to initiate chemical changes, and chemical changes, in their turn, always evolve some mode of force.

EXAMPLES. The mechanical motion communicated by the touch of a feather, will explode nitrogen iodide. A temperature of 150° will cause the explosion of gun-cotton. Electricity, in the electrotype, and light, in the photograph, effect chemical decompositions. We use a match to light our fires; the heat produced by friction causes its combustion, which in its turn, evolves still more heat. All modes of force indeed, may and do result from chemical action.

§ 2. STOICHIOMETRICAL CALCULATIONS.

119. Definition.—Stoichiometry is that department of Chemistry which considers the numerical relations of atoms. All calculations, therefore, which can be made from the atomic weights and volumes are stoichiometrical calculations.

120. Calculations Founded on Weight.—Every atom has its own weight, called the atomic weight. The atomic weight is the smallest portion by weight of any simple or elementary substance—referred to the atom of hydrogen as unity—which can take part in a chemical change.

A molecule being built up of atoms, a molecular weight is the sum of the weights of the atoms of which it is composed. It is also equal to twice the weight of a given volume of a substance in the state of vapor, compared with the same volume of hydrogen.

If the substance be not volatile and cannot be weighed in the state of gas, its molecular weight is the weight of that quantity of it, in its solid condition, which contains the same amount of heat as fourteen parts by weight of lithium at the same temperature.

121. Weight Concerned in Chemical Changes.—Since every chemical change is simply an alteration in the position and association of atoms, every chemical equation which represents such a change, represents it as taking place between definite quantities of matter. An equation expresses not only the fact of chemical reaction between two bodies, but also indicates the quantities by weight concerned in it.

EXAMPLES.—S stands for one atom of sulphur, with an atomic weight of 32. O_3 represents three atoms, or (16×3) 48 parts by weight of oxygen. SO_3 expresses the fact that one atom of sulphur (32) and three atoms of oxygen (48) have united to form a molecule of sulphuric oxide, with a molecular weight of $(48 + 32)$ 80. So $Ca\,(CO_3)$ represents a molecule of calcium carbonate, with a molecular weight of $\overset{Ca}{40} + \overset{C}{12} + (16 \times \overset{O_3}{3}) = 100$. The molecular weight of $K(NO_3)$ is $\overset{K}{39} + \overset{N}{14} + \overset{O_3}{48} = 101$.

So the equation

$$Pb''(NO_3)_2 + Na_2(SO_4) = Pb''(SO_4) + (NaNO_3)_2$$

—one molecule of lead nitrate and one of sodium sulphate, yield one molecule of lead sulphate and two of sodium nitrate—may be read by weight thus :—

$$\begin{array}{ccccccc} 207+(14+48)2 & + & (23\times 2)+32+64 & = & 207+(32+64) & + & (23+14+48)2 \\ Pb''(NO_3)_2 & + & Na_2(SO_4) & = & Pb''(SO_4) & + & (NaNO_3)_2 \\ 331 & + & 142 & = & 303 & + & 170 \end{array}$$

Three-hundred and thirty-one parts of lead nitrate and one-hundred and forty-two parts of sodium sulphate, yield three-hundred and three parts of lead sulphate and one-hundred and seventy parts of sodium nitrate.

122. Calculation of Percentage Composition.—Knowing the molecular weight of any substance, the number of atoms which it contains, and the atomic weight of each, it is easy to calculate its percentage composition; i. e., its composition in 100 parts—the form in which the results of analysis are usually given.

Representing the molecular weight by m, the atomic weight of any constituent by a, the number of atoms of that constituent by n, and its percentage amount by x, then we have evidently, the proportion :—

$$m : an :: 100 : x$$

whence the formula :—

$$x = \frac{an \times 100}{m} \qquad (1)$$

To find therefore, the percentage amount of any constituent in a molecule, we have the following rule :—

Multiply the atomic weight by the number of atoms and this product by 100. Divide the final product by the molecular weight, and the quotient will be the percentage amount of that constituent.

By repeating this process for each atomic constituent, the percentage composition of the molecule may be obtained.

EXAMPLES.—What is the percentage composition of calcium sulphate, $Ca(SO_4)$?

By the formula, the molecule contains of

Calcium, one atom (at. wt. 40) . . .	40
Sulphur, one atom (at. wt. 32) . . .	32
Oxygen, four atoms (at. wt. 16) . .	64
Molecular weight of calcium sulphate,	136

Substituting in the percentage formula, the quantity of

Calcium in 100 parts is $\frac{40 \times 100}{136} = 29 \cdot 41$

Sulphur " " " " $\frac{32 \times 100}{136} = 23 \cdot 53$

Oxygen " " " " $\frac{64 \times 100}{136} = 47 \cdot 06$

100·00

123. Other Problems by this Formula.—In the formula given above (1), the four quantities a, n, m, and x are employed. Any three of them being known, of course the fourth can be found. There are therefore, three more cases to be here considered.

SECOND CASE.—Having the percentage amount of any constituent, its atomic weight, and the molecular weight of the compound given, to find the number of atoms of that constituent in the molecule.

By transposition, formula (1) gives :—

$$n = \frac{mx}{100a} \qquad (2)$$

whence we derive the following rule :—

Multiply the molecular weight by the percentage amount of the given constituent, and divide the product by its atomic weight, multiplied by 100. The quotient is the number of atoms of that constituent in the molecule.

Having obtained in this way the number of each kind of atoms composing a molecule, it is easy to construct the molecular formula.

EXAMPLES.—What is the formula of quartz, its molecular weight being 60, and its percentage composition :—

Silicon	46·67
Oxygen	53·33
	100·00

The atomic weight of silicon is 28; hence by formula (2) the number of atoms of

$$\text{Silicon would be } \frac{60 \times 46{\cdot}67}{100 \times 28} = 1$$

$$\text{Oxygen `` `` } \frac{60 \times 53{\cdot}33}{100 \times 16} = 2$$

The molecular formula of quartz is therefore SiO_2.

THIRD CASE.— Having the percentage composition, the number of atoms of any constituent in the molecule, and the molecular weight, to find the atomic weight of that atomic constituent. From formula (1) by transposition, we obtain :—

$$a = \frac{mx}{100n} \qquad (3)$$

The rule therefore is :—

Multiply the molecular weight by the percentage amount of the constituent whose atomic weight is desired, and divide the product by the number of atoms multiplied by 100. The quotient is the atomic weight required.

EXAMPLES.—The molecular weight of silver nitrate is 170; it contains 63·53 per cent of silver, and has but one atom of silver in a molecule. What is the atomic weight of silver?

Making the necessary substitutions in formula (3) we have $\frac{170 \times 63{\cdot}53}{100 \times 1} = 108$. Hence the atomic weight of silver is 108.

FOURTH CASE.—Having the atomic weight of any constituent, the number of atoms of it in the molecule, and its percentage amount, to find the molecular weight.

By a final transposition of formula (1), we obtain :—

$$m = \frac{an \times 100}{x} \qquad (4)$$

Whence the rule :—

Multiply the atomic weight of the constituent given by the number of its atoms, and this product by 100. Divide the final product by the percentage amount of that constituent and the quotient is the molecular weight.

EXAMPLES.—Salt contains 39·32 per cent of sodium, whose atomic weight is 23. In a molecule of salt there is but one atom of sodium. What is the molecular weight of salt?

By substitution, $\frac{23 \times 1 \times 100}{39{\cdot}32} = 58{\cdot}5$. The molecular weight of salt is therefore 58·5.

Again, ferric oxide contains three atoms of oxygen, or 30 per cent. What is its molecular weight?

By the formula, $\frac{16 \times 3 \times 100}{30} = 160$, the molecular weight.

124. Calculation of an Atomic Group.—In some cases it is desirable to calculate the percentage amount of a group of atoms in any molecule. Formula (1) above given enables us to do this, using a to indicate the weight of the group, and n the number of such groups in the molecule.

EXAMPLES.—Ammonium nitrate, $(NH_4)NO_3$, breaks up under the influence of heat into one molecule of nitrogen oxide, N_2O, and two molecules of water, $(H_2O)_2$. How much nitrogen oxide in 100 parts of ammonium nitrate?

By formula (1) $\frac{an \times 100}{m}$, we have $\frac{44 \times 1 \times 100}{80} = 55$.

Hence ammonium nitrate yields 55 per cent of nitrogen oxide.

125. Other than Percentage Numbers.—Problems often arise which require the quantity of a constituent in less or more than 100 parts. The answers to such problems can of course be obtained by stating the proportion for each problem; but they may be derived also from formula (1) already given, by putting y—the quantity of the constituent—in place of x, and z—the quantity of the compound—in place of 100. The formula then becomes:—

$$y = \frac{an \times z}{m} \qquad (5)$$

Whence the rule:—

Multiply the weight of the constituent contained in one molecule by the weight of the compound given in the problem, and divide this product by the molecular weight. The quotient is the quantity of the constituent required.

EXAMPLES.—How much iodine may be obtained from 236 grams of potassium iodide, the atomic weight of iodine being 127, and the molecular weight of potassium iodide being 166?

By proportion. As 166 parts of potassium iodide give 127 of iodine, it is obvious that the quantity given by 236 parts would be given by the proportion 166 : 236 : : 127 : y. Whence $y = 180{\cdot}5$. Answer, 180·5 grams iodine.

By formula (5). Substituting for an in formula (5), 127, for z, 236, and for m, 166, we have $y = \frac{127 \times 236}{166} = 180{\cdot}5$. Hence 236 grams potassium iodide yield 180·5 grams iodine.

Conversely, if the quantity of the compound necessary to yield a given weight of the constituent be required, we obtain by transposition:—

$$z = \frac{m \times y}{an} \qquad (6)$$

EXAMPLES.—How much potassium iodide would be required to yield 78 grams iodine?

Substituting in formula (6) we have $z = \frac{166 \times 78}{127} = 102$. Answer, 102 grams potassium iodide.

By analysis.—If 166 parts potassium iodide yield 127 of iodine, to yield 1 part of iodine $\frac{166}{127}$ of one part will be required; and to yield 78 parts, 78 times $\frac{166}{127}$ or $\frac{166 \times 78}{127}$, will be required. But this is the precise result given above.

126. Calculation from Equations.—The same principles are applied in calculating the weight of substances entering into, or issuing from, chemical reactions. The reaction is first to be expressed in the form of an equation. The molecular weights of all the substances given are then to be written below their respective formulas. Having now the data, the problems are to be solved by making the following proportion:—

As the molecular weight of the substance given, is to the quantity of it given in the problem, so is the molecular weight of the substance required, to the quantity of it required.

Representing by M, the molecular weight of the substance given, by W, the absolute weight of this substance given in the problem, by m the molecular weight of the substance required, and by w the absolute weight of this substance, then by the above rule we have the proportion $M : W :: m : w$, from which the four following formulas may be derived:—

$$M = \frac{mW}{w} \quad (1) \qquad W = \frac{Mw}{m} \quad (2)$$

$$m = \frac{Mw}{W} \quad (3) \qquad w = \frac{mW}{M} \quad (4)$$

Hence, any three of these quantities being given, it is easy to find the fourth. Four cases thus arise, viz :—

1. Having the absolute quantity of a factor and the quantity of the product yielded by it, as well as the molecular weight of the product ; to find the molecular weight of the factor.

2. Having the molecular weight of both factor and product, to find the quantity of the factor necessary to yield a given weight of the product.

3. Having the quantity of the factor, the quantity of the product, and the molecular weight of the factor ; to find the molecular weight of the product.

4. Having the molecular weight of both factor and product, to find the weight of the product from a given weight of the factor.

EXAMPLES.—Nitric acid is prepared by the action of sulphuric acid upon potassium nitrate, according to the following equation :—

$$K(NO_3) + H_2(SO_4) = H(NO_3) + HK(SO_4)$$
$$101 + 98 = 63 + 136$$

Problem 1st.—125 grams of niter yield 77·97 grams of nitric acid, whose molecular weight is 63 ; what is the molecular weight of potassium nitrate ?

In this problem, m equals 63, W equals 125, and w equals 77·97 ; hence $M = \frac{63 \times 125}{77{\cdot}97} = 101$, Answer.

Problem 2d.—The molecular weight of niter is 101, and that of nitric acid is 63 ; how much niter would be required to yield 77·97 grams nitric acid ?

Here, the quantities being represented as before, we have $W = \frac{101 \times 77{\cdot}97}{63} = 125$ grams, Answer.

Problem 3d.—125 grams of niter yield 77·97 grams nitric acid. The molecular weight of niter is 101 ; what is the molecular weight of nitric acid ?

In this problem, $m = \frac{101 \times 77{\cdot}97}{125} = 63$, Answer.

Problem 4th.—The molecular weight of niter is 101 and that of nitric acid is 63; how much nitric acid would 125 grams of niter yield?

We have $w = \frac{63 \times 125}{101} = 77{\cdot}97$ grams, Answer.

Formulas (2) and (4) are the ones usually employed, since molecular weights may generally be obtained more readily in other ways. These two formulas may be applied to a great variety of problems, as the following examples show.

Examples.—Taking the equation for the production of nitric acid by the action of sulphuric acid on potassium nitrate, above given, the following problems may be worked by formula (2):—

Problem 1st.—How much niter is necessary to yield 36 grams of nitric acid?

$$W = \frac{101 \times 36}{63} = 57{\cdot}7 \text{ grams, Answer.}$$

Problem 2d.—How much sulphuric acid will be required?

Here $M = 98$; hence $W = \frac{98 \times 36}{63} = 56$ grams, Answer.

Problem 3d.—How much hydro-potassium sulphate will be produced?

M in this problem $= 136$; hence $W = \frac{136 \times 36}{63} = 77{\cdot}7$ grams, Answer.

And these problems by formula (4):—

Problem 1st.—How much nitric acid may be produced from 500 grams of potassium nitrate?

$$w = \frac{m\,W}{M} = \frac{63 \times 500}{101} = 311{\cdot}88 \text{ grams, Answer.}$$

Problem 2d.—How much sulphuric acid would be required to decompose 500 grams niter?

Here $m = 98$; hence $w = \frac{98 \times 500}{101} = 485{\cdot}15$ grams, Answer.

Problem 3d.—How much hydro-potassium sulphate would be yielded by the decomposition of 500 grams of potassium nitrate by sulphuric acid?

In this problem, $m = 136$; hence $w = \frac{136 \times 500}{101} = 673{\cdot}27$ grams, Answer.

In all the above problems it has been assumed that each molecule of the factor yielded one of the product. If in any reaction, this is not true, then M and m must represent the sum of the molecular weights expressed in the equation.

127. Volume Calculations from Equations.—Every molecular formula represents two volumes. Hence any equation, composed of such formulas, may be read by volume. From these volumes, calculations may be made, as well as from the weights.

EXAMPLES.—In the equation :—

$$(CO)_2 + O_2 = (CO_2)_2$$

—two molecules carbonous oxide and one molecule of oxygen yield two molecules carbonic di-oxide—the volume-relations may be read thus :—four volumes carbonous oxide and two volumes of oxygen give four volumes of carbonic di-oxide. From these relations the following problems may arise :—

Problem 1st.—How much carbonic di-oxide is formed by the combustion of 1 liter of carbonous oxide ?

As 4 volumes carbonous oxide yield 4 of carbonic di-oxide, 1 volume will yield 1 volume, and 1 liter of course 1 liter, Answer.

Problem 2d.—How much oxygen is needed to convert 2 liters carbonous oxide to carbonic di-oxide ?

Four volumes by the equation require 2 of oxygen ; hence 2 liters will require 1 liter of oxygen, Answer.

Problem 3d.—To form 100 cubic centimeters of carbonic di-oxide how much carbonous oxide must be burned ?

Four volumes carbonic di-oxide require the combustion of four of carbonous oxide ; 100 c. c. will require its own volume therefore, or 100 cubic centimeters, Answer.

128. Relations of Weight to Volume.—It is often necessary to calculate the volume occupied by a given weight of any gas, or the weight of any given volume. The following are the rules :—

1. To determine the volume of any gas, its weight being given: — Divide the weight of the gas given, by the weight of one liter; the quotient is the number of liters.

2. To determine the weight of any given volume of gas:—Multiply the number of liters of gas by the weight of one liter; the product is the weight of the given volume.

EXAMPLES.—1. What volume is occupied by 6·08 grams of oxygen gas?

The weight of one liter of oxygen is 1·43 grams; hence in 6·08 grams there will be as many liters as 1·43 is contained times in 6·08; or 4·25 liters, Answer.

2. What is the weight of 25 liters of nitrogen gas?

One liter of nitrogen gas weighs 1·26 grams. $1·26 \times 25 = 31·5$; hence 25 liters of nitrogen weigh 31·5 grams, Answer.

129. Relation of Volume to Density.—Density being the weight of one volume of any gas—compared with the same volume of hydrogen—and molecular weight being the weight of two volumes, it is evident that the density of any body in the state of gas may be obtained by dividing its molecular weight by two.

Having the density of any gas—which expresses how many times the gas is heavier than hydrogen—the weight of one liter may be readily obtained by multiplying it into the weight of one liter of hydrogen. One liter of hydrogen weighs 0·0896 grams, or 1 crith.

EXAMPLES.

Name.	*Formula.*	*Molecular Weight.*	*Density.*	*Weight of one Liter.* Cal.	Obs.
Oxygen	O_2	32	16	1·4336	1·4298
Nitrogen	N_2	28	14	1·2544	1·2561
Carbonic di-oxide	CO_2	44	22	1·9712	1·9774
Sulphurous oxide	SO_2	64	32	2·8672	.....
Cyanogen	$(CN)_2$	52	26	2·3296	.....

130. Relation of the Hydrogen Unit to the Air Unit.—The word density has been used to indicate the weight of a given volume of gas as compared with hydrogen. The term specific gravity may, in like manner, be used to indicate the weight of a given volume of gas, referred to air as the standard. The density of hydrogen gas is 1, its specific gravity is 0·0693. Hence, by multiplying the density of any gas by 0·0693, its specific gravity may be obtained; and by dividing the specific gravity by 0·0693, the quotient is the density.

EXAMPLES.—What is the specific gravity of chlorine gas?

The molecular weight of chlorine is 71; its density therefore, is $\frac{71}{2}$ or 35·5. 35·5 × 0·0693 = 2·46. Chlorine gas is therefore 2·46 times heavier than air.

Again, the specific gravity of ammonia gas is 0·589. What is its molecular weight?

If the specific gravity is 0·589, its density is 0·589 ÷ 0·0693, or 8·5. Hence its molecular weight is 8·5 × 2, or 17.

131. Reduction of Gaseous Volumes for Pressure.—According to the law of Marriotte, the volume of any gas is inversely, and its density is directly, as the pressure to which it is subjected. Hence the volume of a gas changes with the variations of atmospheric pressure as measured by the barometer; being increased as the barometer falls and diminished as it rises. The normal pressure to which it is common to refer gaseous volumes is 760 millimeters of mercury.

If the volume of a gas under the height H of the barometric column be represented by V, and under any other height H', by V', then by the law above given $V : V' :: H' : H$; whence $VH = V'H'$, or

$$V' = \frac{VH}{H'}$$

Hence, to reduce a given volume of gas to its volume under the normal pressure, we have the following rule :—

Multiply the given volume of gas by the barometric height under which it was measured, and divide the product by 760; the quotient is the true volume.

EXAMPLES.—What is the true volume which 250 cubic centimeters of hydrogen measured at 742 millimeters, would have if measured at 760 m. m.?

By the formula, $V' = V \times \frac{H}{760} = 250 \times \frac{742}{760} = 244$ c. c., Ans.

A certain volume of nitrogen di-oxide gas, under a pressure of 781 m. m., measured 542 c. c. What is its true volume, measured at 760 m. m.?

Substituting in the formula, $V' = 542 \times \frac{781}{760} = 578{\cdot}3$ c. c., Ans.

132. Reduction of Gaseous Volumes for Temperature.—All gases expand or contract by the same amount for the same increase or decrease of temperature. The amount of this expansion—called its co-efficient—is $\frac{1}{273}$ of the volume of the gas at 0°, for every degree Centigrade. For two degrees it would be $\frac{2}{273}$, for three degrees $\frac{3}{273}$, and for t degrees $\frac{t}{273}$. Or, as $\frac{1}{273}$ is equal to 0·003665, one volume of a gas at 0°, becomes 1·003665 volumes at 1°, 1·00733 volumes at 2°, 1 + (·003665 × t) volumes at t°. In general, if V represent the known volume, V' the unknown volume, and t the number of degrees the temperature is raised or lowered, the formula for calculating an increase of volume will be :—

$$V' = V \times (1 + {\cdot}003665t) \qquad (1)$$

And by transposing, the formula by which the volume at a lower temperature can be calculated, is obtained :—

$$V = \frac{V'}{(1 + \cdot003665t)} \qquad (2)$$

EXAMPLES.—1. A gas measures 15 cubic centimeters at 0°; what will it measure at 60°?

Substituting in (1), $V' = 15 \times (1 + 60 \times \cdot003665) = 18{\cdot}298$ c. c., Answer.

2. What will a gas measure at 0°, which, at 100, measures 40·1 c. c.?

Substituting in (2), $V = \frac{40{\cdot}1}{(1 + 100 \times \cdot003665)} = 29{\cdot}345$ c. c., Answer.

3. A gas measures 560 c. c. at 15°; what will it measure at 95°?

Here $t = 95 - 15 = 80$. Hence $V' = 560 \times (1 + 80 \times \cdot003665) = 724{\cdot}2$ c.c., Answer.

EXERCISES.

§ 1.

1. What is molecular stability?
2. What is a chemical reaction? A chemical re-agent?
3. Explain why mass-reactions may be accurately represented by molecular formulas.
4. What is a chemical equation? How is it constructed?
5. Give the rule for writing equations. Illustrate it.
6. Does matter disappear in chemical changes?
7. How are chemical reactions classified? Illustrate.
8. Why does solution favor chemical changes?
9. Give Berthollet's first law. Define precipitate, precipitation.
10. Distinguish the chemical part of this process of precipitation from the physical.
11. Give Berthollet's second law. Illustrate it.
12. How may results be predicted by these laws?
13. If barium chloride and magnesium sulphate be mixed together in solution, will there be a reaction? Lead nitrate and ammonium phosphate? Sodium hydrate and zinc iodide? Write the reaction in each case.
14. In what ways may chemical changes in matter take place? Write a reaction of each kind.

15. Illustrate the strength of chemical attraction.

16. What connection have the physical forces with chemism?

§ 2.

17. What are stoichiometrical calculations?

18. What does a chemical equation represent by weight?

19. Read the following equation by weight:—

$$Sr(NO_3)_2 + HNa_2PO_4 = HSrPO_4 + (NaNO_3)_2$$

20. Deduce the formula for calculating the percentage composition. Give the rule.

21. What is the percentage composition of potassium chlorate? Of sodium carbonate? Of K_3PO_4? Of Zn_2SiO_4?

22. Derive the formula and give the rule, for finding the number of atoms of any constituent in any molecule.

23. Alumina is composed as follows: Aluminum 53·40, Oxygen 46·60 = 100. Its molecular weight is 103; what is its formula?

24. The mineral wollastonite has the following composition: Silicon 24·14, Calcium 34·48, Oxygen 41·38 = 100. Its molecular weight is 116; what is its formula?

25. How is the formula for getting the atomic weight derived? Give the rule.

26. Tin oxide has a molecular weight of 150; it contains one atom of tin, or 78·67 per cent. What is the atomic weight of tin?

27. Magnetic iron oxide contains three atoms of iron; its percentage amount of oxygen is 27·60, and its molecular weight is 232. What is the atomic weight of iron?

28. What is the formula and what the rule for finding the molecular weight?

29. Zinc sulphide contains 67 per cent of zinc, or one atom; the at. wt. of zinc is 65; what is the molecular weight of zinc sulphide?

30. How may the atomic groupings into which a molecule can be broken up, be calculated?

31. The mineral magnesite, $MgCO_3$, is decomposed by heat into MgO and CO_2; what are the percentage amounts of these substances which it contains?

32. Give the formula and rule in cases where other than percentage numbers are required.

33. How much lead may be obtained from 564 kilograms lead sulphide? (At. wt. lead, 207; mol. wt. lead sulphide, 239.)

34. How much calcium phosphate is required to give 356 kilo-

grams phosphorus? (At. wt. of phosphorus is 31; the molecular weight of calcium phosphate is 310.)

35. How may the products of a reaction be calculated from the factors? Give the rule.

36. Derive the four formulas given, and show the class of problems to which each applies.

37. From the following equation :—

$$Zn(NO_3)_2 + K_2CO_3 = ZnCO_3 + (KNO_3)_2$$

calculate the quantity of zinc nitrate required to give 103·17 grams zinc carbonate.

38. How much $ZnCO_3$ may be obtained from 156 grams $Zn(NO_3)_2$?

39. How much K_2CO_3 is needed to decompose 75 gr. $Zn(NO_3)_2$?

40. What quantity of potassium nitrate will result?

41. How much potassium carbonate must be used in order to obtain 54 grams zinc carbonate?

42. How much potassium nitrate will be produced?

43. 156 grams zinc nitrate yield 103·17 grams zinc carbonate (mol. wt. 125); what is the molecular weight of zinc nitrate?

44. 103·17 grams zinc carbonate are obtained from 156 grams of zinc nitrate (mol. wt. 189); what is the mol. wt. of zinc carbonate?

45. Read the following equation by volume :—

$$CH_4 + (O_2)_2 = CO_2 + (H_2O)_2.$$

46. How much oxygen is needed to burn 1 liter of CH_4?

47. What volume of carbonic di-oxide is produced?

48. How much CH_4 is needed to give 1 cubic meter of steam?

49. What volume does a kilogram of oxygen occupy?

50. One liter of CH_4 in burning gives what weight of CO_2?

51. Calculate the weight of one liter of chlorine; of phosphorus; of H_2S; of CO; of PCl_3; of HNO_3.

52. Calculate the specific gravity of nitrogen.

53. The sp. gr. of hydrogen iodide is 4·43; what is its mol. wt.?

54. How is the formula for reducing gaseous volumes to the normal pressure deduced? Give the rule.

55. What is the normal volume of a liter of oxygen measured at 756 m. m.? At 795? At 1140? At 380?

56. What volume would 350 c. c. of ammonia-gas, measured at 74°, have at 0°? At 100°? At 20°?

PART SECOND.

Inorganic Chemistry.

Part Second.

INORGANIC CHEMISTRY.

CHAPTER FIRST.

HYDROGEN.

Symbol H. *Atomic weight* 1. *Equivalence* I. *Density* 1. *Molecular weight* 2. *Molecular Volume* 2. *One liter weighs* 0·0896 *grams (*1 *crith).*

133. History.—Hydrogen was apparently known to Paracelsus in the 16th century. It was first accurately described by Cavendish in 1766, who called it inflammable air. Lavoisier gave it the name hydrogen.

134. Occurrence.—Hydrogen occurs free in certain volcanic gases; Bunsen found that it formed 45 per cent of the gaseous exhalations of Nimarfjall, Iceland. It is shown by the spectroscope to exist in the sun, the fixed stars, and in some of the nebulas. Graham obtained from the Lenarto meteorite—a remarkably pure iron—three times its own volume of this gas.

Combined, hydrogen exists in water, every cubic

centimeter of which contains 1¼ liters; also in petroleum and bitumen, and in all animal and vegetable tissues.

135. Preparation.—Simple molecules are obtained from compound molecules by re-arranging their atoms. For the production of hydrogen this re-arrangement may be effected :—

I. By the action of some physical force; as

(a) Heat.—When melted platinum is dropped into water, both hydrogen and oxygen gases are evolved :—

$$\begin{matrix} H-O-H \\ H-O-H \end{matrix} \quad \text{forming} \quad \begin{matrix} H & O & H \\ | & \| & | \\ H & O & H \end{matrix}$$

or :—

$$(H_2O)_2 = (H_2)_2 + O_2.$$

(b) Electricity.—In the electrolysis of hydrogen compounds.

II. By superior chemical attraction; as in the action of sodium upon water at ordinary temperatures :—

$$\underset{\text{Hydrogen oxide.}}{(H_2O)_2} + \underset{\text{Sodium.}}{Na_2} = \underset{\text{Sodium hydrate.}}{(HNaO)_2} + \underset{\text{Hydrogen.}}{H_2},$$

or of iron and other metals, at a red heat :—

$$\underset{\text{Water.}}{(H_2O)_8} + \underset{\text{Iron.}}{Fe_6} = \underset{\text{Iron oxide.}}{(Fe_3O_4)_2} + \underset{\text{Hydrogen.}}{(H_2)_8}.$$

Or of zinc upon an acid, as sulphuric acid :—

$$\underset{\text{Hydrogen sulphate.}}{H_2(SO_4)} + \underset{\text{Zinc.}}{Zn} = \underset{\text{Zinc sulphate.}}{Zn(SO_4)} + \underset{\text{Hydrogen.}}{H_2},$$

or of magnesium upon a base, as potassic base :—

$$\underset{\text{Potassium hydrate.}}{(HKO)_2} + \underset{\text{Magnesium.}}{Mg} = \underset{\text{Potassio-magnesium oxide.}}{Mg''(OK)_2} + \underset{\text{Hydrogen.}}{H_2}.$$

EXPERIMENTS.—The apparatus by which hydrogen is obtained

by the action of sodium upon water is shown in Fig. 1. It consists of a glass cylinder filled with water, and inverted in water contained in a cistern.—Upon throwing a fragment of sodium upon the water, it rolls about upon the surface with a hissing noise, as a silver-white globule. By means of the wire-gauze cage shown in the figure, this globule may be depressed below the surface, and held beneath the mouth of the glass cylinder. The hydrogen gas set free by the action of the sodium, rises in bubbles into the cylinder, displacing the water. By repeating the process a sufficient number of times, the cylinder may be filled.

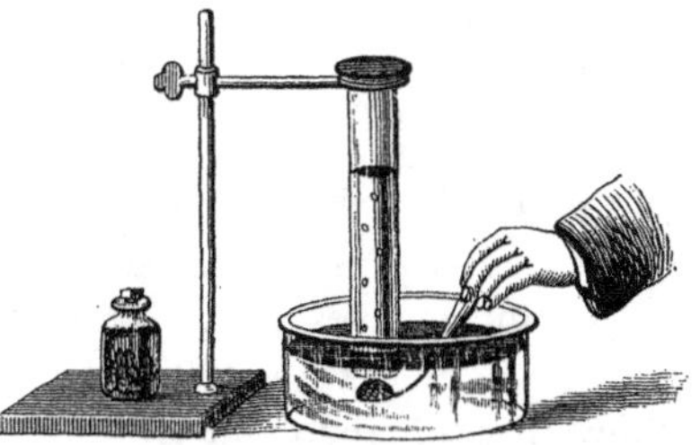

Fig. 1. Preparation of Hydrogen by Sodium.

The usual method of preparing hydrogen is by the action of zinc upon sulphuric acid, for which an apparatus similar to that shown in Fig. 2, may be used. The zinc is placed in the two-necked bottle

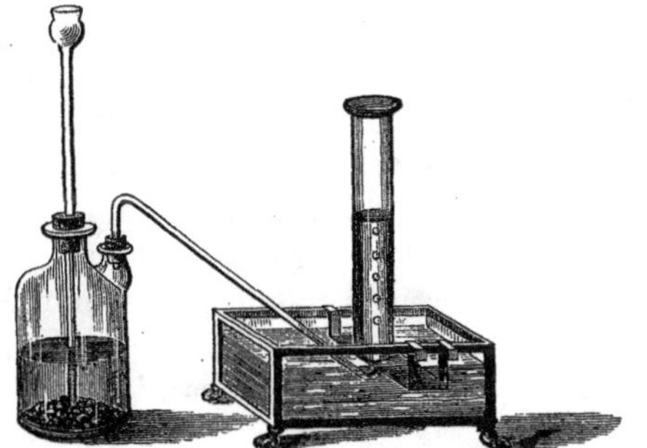

Fig. 2. Preparation of Hydrogen from Zinc and Sulphuric Acid

—in place of which, a wide-mouthed bottle having two holes through the cork, may be substituted; through one of these openings a funnel-tube passes to the bottom of the bottle, and through the other a delivery-tube passes to the water-cistern, terminating beneath an inverted cylinder filled with water, which stands within it. On pouring diluted sulphuric acid—one part of the commercial acid mixed with four parts of water and cooled—through the funnel-tube upon the zinc, effervescence takes place and bubbles escape from the delivery-tube. After allowing time for the air in the bottle to escape, the gas may be collected for use.

136. Properties.—I. PHYSICAL.—Hydrogen is a colorless, odorless, and tasteless gas. It is the lightest form of matter known, being 14·45 times lighter than air, 11,000 times lighter than water, and 240,000 times lighter than platinum. Its molecular weight is therefore smaller than that of any other substance. For this reason, as shown in the section on diffusion, its diffusibility is higher than that of any other gas. Its refractive power on light is remarkable, being 6·614 times that of air. It is soluble to a very slight extent in water, 100 volumes of which dissolve but 1½ of hydrogen. It is a permanent gas, never having been reduced to a liquid by cold or pressure. Owing to its lightness, the velocity of sound in hydrogen is trebled, but its intensity is much enfeebled. Hydrogen is the standard of density, and of molecular weight and volume. 1 liter weighs ·0896 grams, or 1 crith.

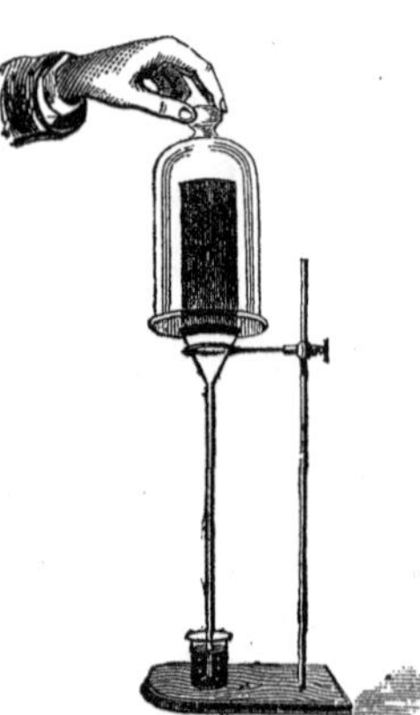

Fig. 3. Diffusion Apparatus.

EXPERIMENTS.—The rapid diffusion of hydrogen gas may be shown very well by the apparatus represented in Fig. 3. A light, unglazed, cylindrical cup of earthen-ware—such as is used in voltaic batteries—is cemented at its open end, to a glass funnel whose stem is prolonged by a slender tube, which dips into colored water. The whole may be supported upon any convenient stand. If now, a bell-glass filled with hydrogen be brought over this earthen cup, the gas diffuses so much more rapidly into the cylinder than the air diffuses out, that an increase of volume takes place within it and the gas bubbles out violently through the water. When the bell-glass is removed, the hydrogen within the cylinder being now in excess, diffuses so rapidly outward as to produce a partial vacuum, so that the colored water rises half a meter or more in the tube.

The levity of hydrogen may be shown by using the gas to inflate soap-bubbles. When detached from the pipe, they rise rapidly. Any bag made of thin tissue, such as collodion or varnished paper, may be filled with hydrogen and will then rise like a balloon.

The curious effect of hydrogen upon sound may be illustrated by placing in a large bell-glass, suspended mouth downward, one of the squeaking images used as toys for children. As the image passes up into the gas, the sound is observed to be greatly enfeebled, and altered considerably in character. The same fact may be shown with the human voice by filling the lungs with the gas, and then speaking. Especial care should be taken however, to have the gas for this purpose made from pure materials; for, although the lungs may be filled once with pure hydrogen gas without injury, yet it is liable to contain impurities which may produce serious results.

II. CHEMICAL.—Hydrogen gas is combustible; that is, when heated to a certain degree—about 500°—it is capable of combining with the oxygen of the air with the evolution of light and heat. The flame of burning hydrogen is pale, and, under the atmospheric pressure, is scarcely luminous; though it becomes bright if the pressure be increased. The heat evolved by it is very great; one gram of hydrogen in burning produces heat sufficient to raise 34,462 grams of water from 0° to 1°; i. e., 34,462 heat-units. It does not support combustion or respiration; a lighted candle placed in it is extinguished and an animal loses his life when confined in it. It is the standard of atomic weight and of equivalence.

EXPERIMENTS.—A cylinder full of gas—collected as shown in Fig. 2, for example—may be inverted, and a lighted taper applied to its mouth. The hydrogen takes fire with a slight explosion and burns with its characteristic flame. If the jar be held mouth downward, and the candle be passed up into it, as shown in Fig. 4, the gas takes fire and burns quietly at the open end, while the flame of the candle, as it passes into the gas, is extinguished, but may be relighted again from the burning hydrogen as it is withdrawn.

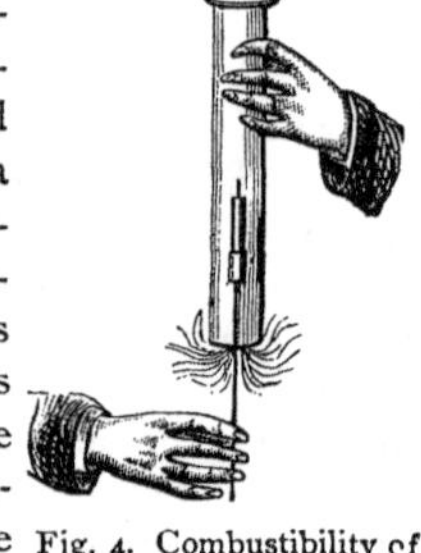

Fig. 4. Combustibility of Hydrogen.

The combustibility, and at the same time the levity, of hydrogen may be shown by covering a bell-glass of this gas with a glass plate, and holding it, mouth upward, beneath a lighted candle 6 or 8 inches distant. On removing the plate, the gas rises from the bell, comes in contact with the flame and takes fire with a slight explosion. The same fact is shown by pouring a bell-glass full of this gas upward into an empty bell, testing each, after the experiment, with a lighted candle. If the soap bubbles above mentioned be touched with a lighted taper as they ascend, Fig. 5, they take fire and burn with a slightly yellow flame.

Fig. 5. Lighting a soap-bubble filled with Hydrogen.

Water is the sole product of the combustion of hydrogen. Hence its name, from ὕδωρ and γεννάω, water-former.

When burned from a jet, as shown in Fig. 6—being previously dried by passing it through a tube containing calcium chloride—the flame of hydrogen, though pale, is very hot, and will raise a small coil of fine platinum wire placed within it to a white heat. On holding a cold and dry bell-glass over this flame, it is at once dimmed with the moisture, and if the experiment be sufficiently long-continued the water produced will run down the sides of the bell-glass in drops.

Within a few years, Graham has pointed out the fact

that hydrogen is capable of being absorbed or occluded by many metals, at temperatures more or less elevated. Of these metals, palladium is the most remarkable, being able to take up over nine-hundred times its volume of hydrogen at ordinary temperatures, forming a white metallic solid, containing its constituents in ratios nearly atomic. Graham maintained that the hydrogen in this substance is a solid metal, with a density about 2, and analogous in many respects to magnesium; that it has a certain amount of tenacity, conducts electricity readily, and is magnetic. He therefore proposed for it the name hydrogenium.

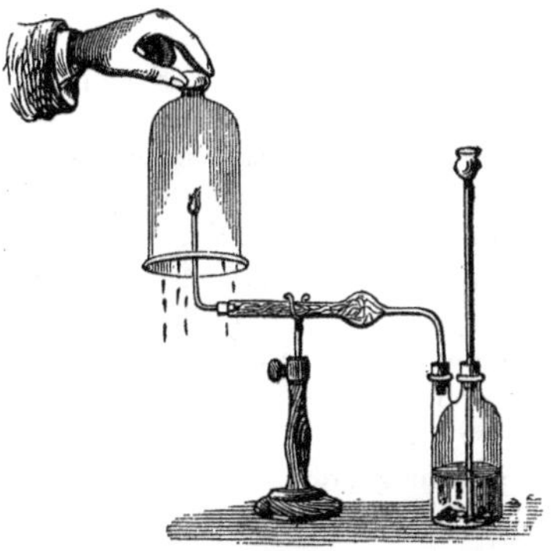

Fig. 6. Water from the combustion of Hydrogen.

137. Uses.—On account of its lightness, it has been used to fill balloons for military and other purposes. The amount which a balloon will carry up, i. e., its ascensional power, is the difference between the weight of the balloon itself with its contained hydrogen, and that of an equal volume of air. A liter of hydrogen gas has an ascensional force of 1·2 grams.

Hydrogen is used also in the arts as a heating material, on account of the high temperature developed by its combustion.

EXERCISES.

1. Mention some substances which contain hydrogen.

2. Write the equation which expresses the preparation of hydrogen by the action of potassium upon water.

3. Give the reaction which takes place when iron acts on sulphuric acid.

4. Ten grams of water will give how many grams of hydrogen when decomposed by heat? By the action of sodium?

5. How many cubic centimeters in each case?

6. How many grams of sodium will be required?

7. Ten liters of hydrogen are desired; how many grams of zinc are necessary to furnish this quantity? How many grams of iron?

8. Twenty grams magnesium will yield how many liters of hydrogen? How many grams of potassium hydrate must be employed?

9. What will hydrogen cost per cubic meter, when made with iron costing 7 cents and sulphuric acid costing 15 cents per kilogram? When made with zinc costing 22 cents per kilogram?

10. What volume of hydrogen does one liter of water contain? One liter of water-vapor or steam?

11. What volume does 0·423 grams of hydrogen occupy at 0°? At 15°? At 100°?

12. Calculate the specific gravity of hydrogen from its density.

13. To what temperature must air be raised to have the density of hydrogen at 0°?

14. Under what barometric pressure has air the density of hydrogen?

15. From what does the name hydrogen come?

16. What is a unit of heat? How many heat-units does hydrogen produce in its combustion?

17. How many grams of hydrogen must be burned to raise 50 kilograms of water from 0° to 10°?

18. What must be the diameter of a spherical balloon which, when filled with hydrogen, will have an ascensional force of 80 kilograms; the balloon itself weighing 30 kilograms?

19. What will it cost to inflate the above balloon, if the hydrogen be prepared by the use of sodium at ten dollars the kilogram?

CHAPTER SECOND.

NEGATIVE MONADS.

§ 1. Chlorine.

Symbol Cl. *Atomic weight* 35·5. *Equivalence* I, III, V, *and* VII. *Density* 35·5. *Molecular weight* 71. *Molecular volume* 2. 1 *liter weighs* 3·17 *grams* (35·5 *criths*).

138. History.—Chlorine was first obtained by Scheele in 1774, and called dephlogisticated muriatic acid; a name afterward changed to oxymuriatic acid by Berthollet. In 1809, Gay-Lussac and Thenard suggested its elementary character, which was established by Davy in 1810, who gave it the name it bears.

139. Occurrence.—Chlorine never occurs free in nature. In combination with sodium, magnesium, potassium and calcium, it exists abundantly in saline springs, and also in sea-water, every liter of which contains 5 liters of chlorine. Sodium chloride or salt, exists also in the solid form in the earth, forming vast deposits, many of which are mined.

140. Preparation.—Chlorine may be prepared:—

I. By the action of heat or of electricity upon chlorides.

$$\underset{\textit{Platinic chloride.}}{PtCl_4} = \underset{\textit{Platinous chloride.}}{PtCl_2} + \underset{\textit{Chlorine.}}{Cl_2}.$$

II. By the superior chemism of oxygen; as when hydrogen chloride acts upon manganese di-oxide:—

$$(HCl)_4 + MnO_2 = MnCl_2 + (H_2O)_2 + Cl_2.$$

Hydrogen chloride. *Manganese di-oxide* *Manganous chloride.* *Water.* *Chlorine.*

Or, when sodium chloride, sulphuric acid and manganese di-oxide are heated together :—

$$(NaCl)_2 + (H_2(SO_4))_2 + MnO_2 =$$

Sodium chloride. *Sulphuric acid.* *Manganese di-oxide.*

$$Mn(SO_4) + Na_2(SO_4) + (H_2O)_2 + Cl_2.$$

Manganous sulphate. *Sodium sulphate.* *Water.* *Chlorine.*

EXPERIMENTS.—The apparatus employed for preparing chlorine is shown in Fig. 7. The materials are placed in a flask which

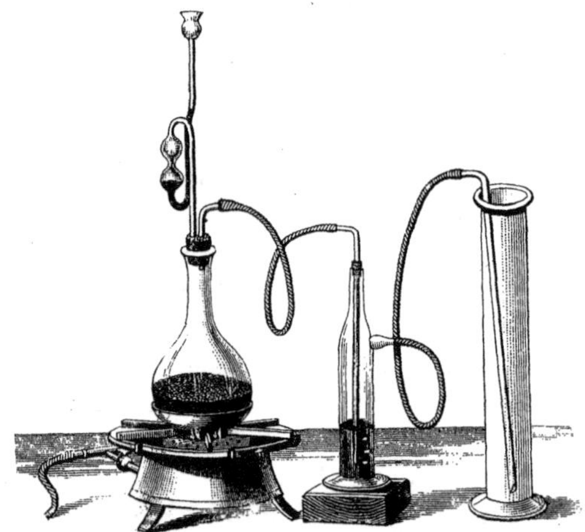

Fig. 7. Preparation of Chlorine.

stands in sand contained in a thin iron cup, upon the gas furnace. Through the cork of this flask two tubes pass, one for the delivery of the gas, the other a safety-tube. This safety-tube is a funnel-tube bent twice upon itself, upon the recurved portion of which are two bulbs. When any liquid is poured into the funnel, a portion remains in the bend and acts as a valve to prevent the escape of the gas. Should the pressure within be increased, the gas will force the liquid up into the funnel and escape through it in bubbles;

should it be diminished, the outside pressure will force the liquid into the bulbs and air will enter, thus avoiding accident. To the delivery tube is attached, by means of a rubber tube, a bottle containing sulphuric acid, to the bottom of which this delivery tube passes, and through which the gas is made to bubble, in order to dry it. From this drying-bottle it passes through a long glass tube to the bottom of the cylindrical gas-jar, where, being heavier than air, the gas gradually collects. When full, a fact easily ascertained from the green color of the gas, the mouth of the cylinder is closed with a glass plate smeared with a little tallow. For every liter of chlorine gas, 8 grams of manganese di-oxide and 20 grams of hydrochloric acid (of commercial strength, sp. gr. about 1·16) are required. The acid is placed in the flask first, the di-oxide is then added, and the whole agitated. The evolution goes on for a time without heat, but to complete the operation the gas beneath must be lighted.

141. Properties.—I. PHYSICAL. Chlorine is a yellowish-green gas—its name coming from χλωρός, yellowish-green—of a peculiar suffocating odor and astringent taste. It is totally irrespirable, producing coughing, even when very dilute, and in larger quantity, inflammation of the air-passages. Its specific gravity is 2·46; it is therefore nearly two and one half times heavier than air. Under a pressure of four atmospheres at the ordinary temperature, or when subjected without pressure, to a cold of −40°, it is condensed to a dark yellow liquid of specific gravity 1·38, which does not solidify even when cooled to −110°. It is quite soluble in water, one volume of which at 11°, dissolves nearly 3 volumes of chlorine gas, forming a solution which possesses essentially the properties of the gas itself. Cooled to 0°, a definite molecular compound crystallizes out, which contains, to every molecule of chlorine, 10 molecules of water.

II. CHEMICAL.—Chlorine has an exceedingly strong attraction for other substances. It combines directly with all the elements except oxygen, nitrogen, and carbon. When finely divided copper, antimony, or arsenic, is placed in the gas, it combines with it, with the evolutions of light and heat to form a chloride. Phosphorus at ordinary temperatures, and sodium at more elevated ones, burn in chlorine spontaneously, forming phosphoric and sodium chlorides. Its attraction for hydrogen is specially strong, the two gases exploding violently when mixed together and exposed to sunlight, or on the approach of a flame. In an atmosphere of hydrogen, chlorine gas burns freely, as hydrogen burns freely in one of chlorine. The heat of the combustion of hydrogen in chlorine is less than in oxygen, being but 23,783 units. Chlorine does not burn in the air at any temperature, owing to its slight attraction for oxygen.

ALLOTROPISM.—Chlorine is capable of existing in two states or conditions, the one active, the other passive. The passive condition is the one obtained when the chlorine is prepared in the dark; when prepared in full daylight it becomes exceedingly active, capable of effecting unions not before directly possible. Chlorine prepared in the dark, may be mixed with hydrogen, without combination taking place. But place the mixture in the full sunlight and it at once explodes.

The existence of an element in two conditions, in one of which it has different properties from those exhibited by the other, is called allotropism. The substance is said to exist in two allotropic states. It is probable that allotropism is due only to differences in molecular atomicity.

One of the most noticeable properties of chlorine is its bleaching power, due to its attraction for hydrogen. Water is decomposed by active chlorine, and the oxygen which is thereby set free, destroys the vegetable coloring matter. Mineral colors in general, are unaffected by chlorine. Its tendency to unite with hydrogen and thus to destroy foul smelling gases, of which this hydrogen is a constituent, is the cause of its value as a disinfecting agent.

EXPERIMENTS.—The action of chlorine upon the metals may be shown by dropping into a jar of the gas, thin leaves of copper or bronze-leaf, or shaking into it some powdered arsenic or antimony. The metals burn spontaneously and vividly as they enter the gas. Phosphorus, introduced in a combustion-spoon—a cup attached to the end of a bent wire—is instantly inflamed. Sodium, if previously heated to redness, burns brilliantly in a jar of chlorine.

Combustion Spoon.

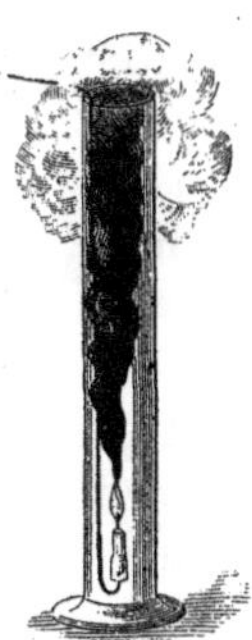

Fig. 8. Candle burning in Chlorine.

A lighted candle lowered into the gas, burns with a smoky flame at first, as shown in Fig. 8, but is soon extinguished. The chlorine takes the hydrogen of which the wax is composed, and the carbon is set free in the form of smoke. A more striking way of showing the relative attractions of chlorine for hydrogen and for carbon is shown in Fig. 9. A jar of pure chlorine has thrust into it a piece of thin tissue paper, previously moistened with *warm* oil of turpentine. The chlorine seizes the hydrogen of the turpentine, evolving so much heat in the combination, that the whole takes fire, evolving dense clouds of smoke.

Fig. 9. Turpentine in Chlorine.

Either the gas, or its solution in water, may be used to show its bleaching power. Pieces of print, having various designs upon them in color, when moistened and placed in contact with the chlorine, will have their colors discharged. A piece of paper covered with characters, partly written, and partly printed, loses entirely all the writing upon it when placed in chlorine; the writing-ink being attacked, while the printing ink, made of carbon, is unaffected.

142. Tests.—Free chlorine may be detected readily by its odor, and, if pure, by its color; by its bleaching action upon indigo, and by the dense white fumes which it gives with ammonia. In solution, it is detected by its power of dissolving gold leaf. In combination, it yields with solutions of silver salts, a white precipitate of silver chloride, insoluble in nitric acid.

143. Uses.—Chlorine is used very largely in the arts for the preparation of the so-called chloride of lime, the form in which this gas is made available as a bleaching agent. The process employed on the large scale is the second given in a former section. Salt and sulphuric acid, together with manganese di-oxide, or sometimes nitric acid, are heated together in large leaden vessels, and the gas as produced is conducted into long, low, stone chambers, upon the floor of which dry slacked lime is placed for the purpose of absorbing it.

COMPOUNDS OF CHLORINE WITH HYDROGEN.

HYDROGEN CHLORIDE. — *Formula* HCl. *Molecular weight* 36·5. *Molecular volume* 2. *Density* 18·25. 1 *liter weighs* 1·63 *grams.* (18·25 *criths*).

144. History.—Only one compound of chlorine and hydrogen is known; this is hydrogen chloride, or as it is more frequently called, hydrochloric acid. It was known to the alchemists under the name "spirit of salt." Glauber in the 17th century gave it the name muriatic acid, from the Latin muria, brine. The pure gas was first obtained by Priestley in 1772, though it was not till 1810 that its true composition was ascertained by Davy.

145. Preparation.—Since chlorine and hydrogen are both monads, and their molecules are both di-atomic, it follows that they unite in equal volumes. Hydrochloric acid may be formed therefore by the direct union of equal volumes of its constituents, according to the equation:—

$$H_2 + Cl_2 = (HCl)_2.$$

This equation also declares that the volume of the hydrogen chloride is the same as that occupied by the hydrogen and chlorine which formed it.

Fig. 10. Direct union of Hydrogen and Chlorine.

EXPERIMENTS.—If a suitable jar be filled, one half with hydrogen, the other half with chlorine, and a flame be applied to the open mouth as shown in Fig. 10, a smart explosion takes place, and white fumes of hydrochloric acid are formed. In this experiment it is well to wrap a towel about the cylinder, to prevent the pieces from flying in case of breakage. If the glass be strong enough, the mixture of gases may be exploded by exposing the jar to sunlight. On opening it afterward, with the mouth under mercury, none will enter, thus showing that the volume of the hydrochloric acid produced is the same as that of the chlorine and hydrogen before union.

II. The second method of preparing hydrochloric acid, and the one usually employed, is by the action of sulphuric acid upon chlorides, generally sodium chloride, or salt. The reaction is :—

$$(NaCl)_2 + H_2(SO_4) = Na_2(SO_4) + (HCl)_2.$$

Sodium chloride. Hydrogen sulphate. Sodium sulphate. Hydrogen chloride.

By passing the gas thus set free through water so long as it is absorbed, the liquid acid is obtained.

EXPERIMENTS.—For the preparation of the gas from salt and sulphuric acid, the apparatus shown in Fig 11 may be employed.

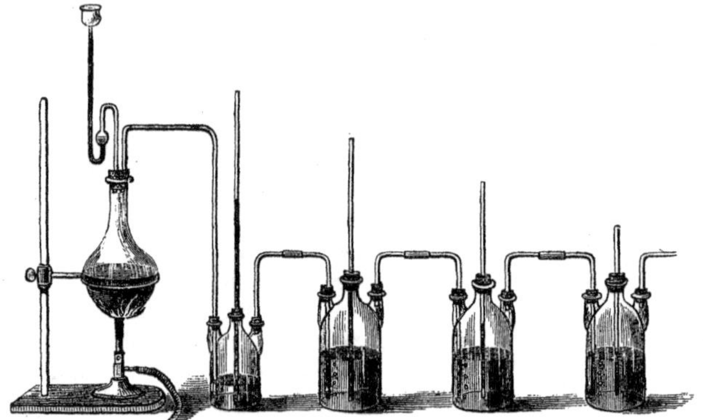

Fig. 11. Preparation of Hydrochloric Acid.

The salt is placed in the flask, and upon it is poured, through the safety tube, twice its weight of sulphuric acid, previously diluted with one-fourth its volume of water. Upon applying a gentle heat, the gas is copiously evolved, and may be collected either over mercury or by displacement.

If it is desired to obtain the liquid acid, the gas is passed through water contained in the series of bottles shown in the figure, called Woulfe's bottles. The first bottle, which is smaller than the others, contains water to wash the gas, which then passes into the larger bottles, charging the water in each, in succession.

146. Properties.—Hydrogen chloride is a colorless, pungent, acid gas, which fumes strongly in the air, is irrespirable, and extinguishes flame. When subjected to a pressure of 40 atmospheres at 10°, or of 2 atmospheres at −70°, it is condensed to a colorless limpid liquid having a specific gravity of 1·27. It is remarkably soluble in water, one volume of which dissolves 450 volumes of this gas at 15°, forming the liquid acid. This acid has a specific gravity of 1·21; when heated it evolves hydrochloric acid gas, until, under the ordinary atmospheric pressure, the solution contains 20·22 per cent of the acid; then the boiling point remains stationary at 110°, and the liquid distills over unchanged.

EXPERIMENTS.—The solubility of hydrogen chloride in water, and at the same time its acidity, may be shown by removing the stopper of a tall cylinder filled with the dry gas, beneath the surface of water colored with blue litmus solution. On agitating the

Fig. 12. Commercial Preparation of HCl.

vessel a little, the water enters as if into a vacuum, the cylinder being not unfrequently broken.

147. Uses.—Hydrochloric acid is manufactured on an immense scale in the arts, chiefly as a waste product in the soda industry. The salt is treated with sulphuric acid in cast iron cylinders placed in a furnace, as shown in Fig. 12, and the gas as evolved, is condensed in water contained in large Woulfe's bottles made of earthenware. It is used for various minor purposes in the chemical arts.

§ 2. Bromine, Iodine, and Fluorine.

Bromine.—*Symbol* Br. *Atomic weight* 80. *Equivalence* I, V, *and* VII. *Specific gravity* 3·187 *at* 0°. *Density of vapor* 80. *Molecular weight* 160. *Molecular volume* 2. 1 *liter of bromine-vapor weighs* 7·15 *grams* (80 *criths*).

148. History.—Bromine was discovered in the water of the Mediterranean sea by Balard, in 1826. On account of its disagreeable odor, he gave it the name bromine, from βρῶμος, stench.

149. Preparation.—On evaporating the water of many saline springs, the salt crystallizes out, and there is left a solution of the more soluble salts—called the mother-liquor or bittern—which is rich in bromides. By heating this bittern with manganese di-oxide and sulphuric acid, the chlorides in it are decomposed, yielding chlorine, which, in its turn, sets the bromine free from the bromides. The vapors of the bromine are led into a cooled receiver, where they condense into a liquid.

150. Properties.—Bromine is a heavy, dark brownish-red liquid, of a disagreeable odor, somewhat recalling that of chlorine. At a temperature of 63° it boils, and is converted into a deep-red vapor which is five and one half times heavier than air. Cooled to −22°, it becomes a dark lead-gray crystalline solid, with a metallic luster. It is but slightly soluble in water, thirty-three parts of which dissolve, at 15°, but one part of bromine.

Chemically it is similar to chlorine, but less active. Several of the metals burn in its vapor, and it exerts a decided bleaching action. It is an active corrosive poison.

Bromine colors starch yellow; and a bromide in aqueous solution precipitates silver from its solutions, as yellow silver bromide. It is used principally in photography.

IODINE.—*Symbol* I. *Atomic weight* 127. *Equivalence* I, III, V, *and* VII. *Specific gravity* 4·95. *Vapor-density* 127. *Molecular weight* 254. *Molecular volume* 2. 1 *liter of iodine-vapor weighs* 11·37 *grams (*127 *criths).*

151. History.—On the coasts of Scotland and Normandy large masses of sea-weeds were formerly burned in order to extract the soda which they contain. The semi-fused ash—called kelp or varec—was dissolved in water and the soda salts crystallized out. It was in the mother-liquor thus obtained, that iodine was first discovered by Courtois, in 1811. Its elementary character was determined by Gay-Lussac in 1813. Its name is from ἰώδης, violet-colored, in allusion to the beautiful color of its vapor.

152. Preparation.—It is prepared commercially in the same way as bromine, by distilling the mother-liquors just mentioned with manganese di-oxide and sulphuric acid, and condensing the vapors.

153. Properties.—Iodine is a bluish-black solid, having a metallic luster, and crystallizing in rhombic scales, or, not infrequently, in well-defined ortho-rhombic octahedrons, Fig. 13. Heated to 107° it melts, and at 180° boils, evolving a dense, magnificent violet vapor, which is 8·72 times heavier than air, and is therefore, the heaviest vapor known. It is only slightly soluble in water, one part of iodine requiring 7000 parts of water at ordinary temperatures, to dissolve it. It dissolves readily in alcohol, ether, carbon disulphide and chloroform; and also in aqueous solutions of potassium iodide.

Fig. 13. Crystals of Iodine.

In chemical activity it is third in the series, being next to bromine. It bleaches but faintly, if at all, in full sunlight, but combines directly with the metals forming iodides. It stains the skin yellow, but is not an active poison.

Starch is a characteristic reagent for free iodine. It strikes with it a deep blue color, which is so intense that one part of iodine may be detected by it in 300,000 parts of water. The most delicate test for iodine is the purple-red color it produces when dissolved in carbon disulphide. One part of iodine in 1,000,000 parts of water may be detected in this way. Soluble iodides produce characteristic precipitates with mercurous, mercuric, and lead salts.

EXPERIMENTS.—The process of obtaining iodine may be illustrated by adding to a solution of potassium iodide contained in a test-tube, a few drops of chlorine-water. The liquid becomes brown in color from the iodine set free, and upon agitating it with ether, the iodine is dissolved by the ether and forms a dark brown layer above, while the solution below is colorless. Upon pouring off the ether and agitating it with a little solution of potassium hydrate, potassium iodide is formed and the ether is decolorized.

The fumes of iodine are conveniently exhibited by throwing some iodine upon a heated brick and then covering the whole with a large bell-glass. To show its reaction with starch, add a drop of the alcoholic solution to a very dilute solution of starch contained in a tall jar. The blue color, upon agitation, will penetrate the entire mass. This experiment may be modified by adding a few drops of potassium iodide to a dilute solution of starch, and then a few drops of chlorine water. The iodine does not color the starch until set free by the chlorine, when the blue appears.

Add a few drops of potassium iodide to some water contained in each of three tall test-glasses. Upon dropping into the first a little of a solution of lead acetate, a brilliant yellow precipitate of lead iodide is obtained. The second, treated with a few drops of mercurous nitrate gives a bright yellowish-green precipitate of mercurous iodide. While the third, upon the addition of a little mercuric chloride, gives scarlet mercuric iodide.

154. Uses.—Iodine is largely used in medicine, both free and in combination. It is particularly serviceable in glandular affections. It is also used extensively in photography.

FLUORINE.—*Symbol* F. *Atomic weight* 19. *Equivalence* 1. *Molecular weight* 38 *(?) Molecular volume* 2 *(?) Density* 19 *(?)* 1 *liter weighs* 1·7 *grams (*19 *criths.) (?)*

155. Preparation and Properties.—The mineral known as fluorite, fluor or Derbyshire spar, is a compound of fluorine with calcium. By means of this calcium fluoride, silver fluoride may be prepared, and by the action of dry iodine upon this, a colorless gas has been obtained which is supposed to be free fluorine. Its name comes from that of the mineral just mentioned. Fluorite comes from the Latin fluo, I flow, because it is used as a flux in the reduction of metals.

HYDROBROMIC, HYDRIODIC AND HYDROFLUORIC ACIDS.

156. Hydrogen Bromide or Hydrobromic acid. — *Formula* HBr. *Molecular weight* 81. *Molecular volume* 2. *Density* 40·5

Hydrogen bromide may be obtained by acting upon phosphorus with bromine, in presence of water. The phosphorus and the bromine first unite to form phosphoric bromide according to the following reaction :—

$$P_4 + (Br_2)_{10} = (PBr_5)_4$$

which as fast as formed, is decomposed into phosphoric acid and hydrobromic acid by the water present :—

$$PBr_5 + (H_2O)_4 = H_3PO_4 + (HBr)_5.$$

Hydrobromic acid is a colorless acid gas, fuming strongly in moist air, and very soluble in water. Cooled to $-73°$ it becomes a liquid, and this at a little lower temperature, freezes.

157. Hydrogen Iodide or Hydriodic acid. — *Formula* HI. *Molecular weight* 128. *Molecular volume* 2. *Density* 64.

Hydrogen iodide may be readily prepared by heating together potassium iodide, iodine, and phosphorus, in presence of water. The reaction is analogous to that above given for hydrogen bromide :—

$$\underset{\text{Potassium iodide.}}{(KI)_4} + \underset{\text{Phosphorus.}}{P_2} + \underset{\text{Iodine.}}{(I_2)_5} + \underset{\text{Water.}}{(H_2O)_8} =$$

$$\underset{\text{Hydriodic acid.}}{(HI)_{14}} + \underset{\text{Hydro-di-potassium phosphate.}}{(HK_2(PO_4))_2}$$

Hydriodic acid also, is a colorless gas which has an acid reaction and fumes in the air. It is easily condensed to a liquid by a pressure of four atmospheres at 0°, and this liquid, cooled to −51°, freezes to a clear ice-like solid. It is as soluble in water as hydrogen chloride, yielding a solution of specific gravity 1·7, which boils at 127°, and contains 57 per cent of hydrogen iodide. This solution is much used as a reagent in organic chemistry.

158. Hydrofluoric acid. — *Formula* HF. *Molecular weight* 20. *Molecular volume* 2. *Density* 10.

Hydrogen fluoride is generally prepared by the action of sulphuric acid upon calcium fluoride; calcium sulphate and hydrogen fluoride resulting :—

$$CaF_2 + H_2(SO_4) = Ca(SO_4) + (HF)_2.$$

The process must be conducted in a vessel of lead or of platinum, as the gas readily attacks glass.

Hydrofluoric acid is a colorless gas, which fumes in the air, and reacts strongly acid. The anhydrous gas does not condense to a liquid at −12°; though as obtained by the above process, it is a liquid having a specific gravity of 1·06, owing to the presence of moisture. The strong acid corrodes the skin powerfully, giving rise to painful ulcers; and the fumes, if inhaled, produce serious irritation of the lungs.

This substance is distinguished from all others by its remarkable property of attacking glass. With the silicon of the glass the fluorine combines to form a gaseous silicon fluoride. This acid is used extensively in the arts for etching glass, the highly ornamented door-lights now so common being prepared by this agent. Used as a liquid the etched surface is left smooth; but when the gas is applied, the surface remains rough.

EXPERIMENT.—Place some finely pulverized fluor-spar or cryolite in a dish of lead, or preferably of platinum, Fig. 14, and pour upon it some strong sulphuric acid. Cover the dish with a glass plate, upon the lower side of which is a thin layer of wax, through which some characters have been drawn with a fine point. Place the whole on a suitable stand and heat very gently for a few minutes. If now the wax be removed from the plate, the device drawn will be found etched upon the glass.

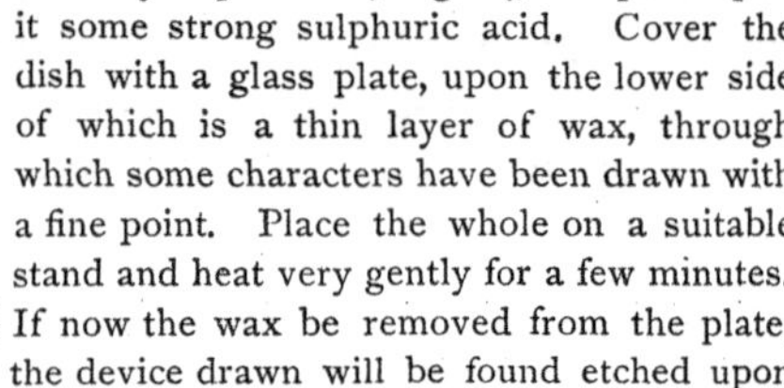

Fig. 14. Etching by HF.

§ 3. PROPERTIES OF THE HALOGEN GROUP.

159. Chlorine, bromine, and iodine constitute a closely-allied group of elements. Even in their physical properties, there is a remarkable progression observable. As to physical state, chlorine is a gas, bromine a liquid, iodine a solid; as to color, chlorine is yellowish-green, bromine-vapor is brown, iodine-vapor purple; as to density, chlorine comes first, then bromine, then iodine. All of them exist in the solid, liquid, and gaseous forms, and change from one to the other at temperatures not far apart.

The same is true of their chemical properties. The atomic weight of bromine, 80, is nearly a mean between that of chlorine and iodine, $\frac{127 + 35.5}{2} = 81.25$. Chlorine is more active than bromine, and bromine more than iodine. Indeed, as is frequently the case with allied elements, the chemism seems to vary inversely as the atomic weight.

Moreover, the hydrogen compounds of these elements are similarly constituted and exhibit a similar gradation in properties. Their binary compounds with potassium and sodium all resemble sea-salt; hence these compounds are frequently called haloid salts, and the elements halogens.

EXERCISES.

§ 1.

1. In what compounds does chlorine occur in nature?

2. What volume does 32 grams of chlorine occupy?

3. What is the weight of 2356 cubic centimeters?

4. How many grams of platinic chloride are required to give 25 grams of chlorine? To give 500 cubic centimeters?

5. How many liters of chlorine from 20 grams of hydrochloric acid gas? How much manganese di-oxide is needed?

6. Liquid hydrochloric acid of sp. gr. 1·20 contains 41 per cent of HCl; one liter will yield how many liters of chlorine?

7. Commercial manganese di-oxide is seldom pure; what per cent of MnO_2 does a sample contain, 30 grams of which heated with HCl, gives 12·04 grams chlorine?

8. One kilogram of salt contains how many grams of chlorine?

9. One cubic centimeter salt contains how many cubic centimeters of chlorine? (Sp. gr. of salt 2·15).

10. 150 liters of chlorine, at 15°, and under 742 m.m. pressure, are required; how many grams of salt, of sulphuric acid, and of manganese di-oxide, are necessary?

11. Give the formulas and names of the compounds of chlorine with Ag', Cu'', Co'', Sb''', B''', Sn^{IV}, Ti^{IV}, P^{V}, $(Fe_2)^{VI}$.

12. Calculate the percentage composition of silver chloride.

13. If ten c. c. of hydrogen diffuse into an atmosphere of chlorine

in one minute, how much chlorine will diffuse into the hydrogen in the same time?

14. One cubic meter of HCl contains what weight of Cl?

15. What volume of Cl will convert one liter of H into HCl?

16. Calculate the sp. gr. of HCl from its density.

17. What volume has one kilogram of HCl gas?

18. One liter of HCl passed over heated iron, yields what volume of hydrogen?

19. Write the reaction which takes place in replacing the sodium in salt by hydrogen.

20. How many kilograms of HCl gas may be obtained from 28 kilograms of salt? How many liters?

21. How much sulphuric acid would be required to decompose it? How much $Na_2(SO_4)$ would be obtained?

22. One liter of liquid acid, sp. gr. 1·20 contains how many grams of HCl? How many liters?

23. How many kilograms of salt and of sulphuric acid are required to produce 150 kilograms of liquid HCl, containing 24·24 per cent of the gas?

24. What weight of potassium hydrate is required to neutralize one liter of hydrochloric acid gas?

25. Ten grams of the liquid acid precipitated 2·26 grams silver chloride from a solution of silver; what percentage of HCl did the solution contain?

§ 2.

26. What is the weight of one c. c. bromine-vapor at 150° and 760 m. m.?

27. Fifty c. c. of HBr decomposed by Na, gives what vol. of H?

28. What volume of HBr will 10 grams of Br give?

29. If a liter of HI and one of chlorine be mixed together, what will be the reaction? What will be the resulting gaseous volume and what its composition?

30. Show that the liquid hydriodic acid which contains 57 per cent of HI, is not a definite hydrate.

31. What volume does 100 grams of HF occupy?

32. What volume of H is contained in 250 c. c. of HF?

CHAPTER THIRD.

NEGATIVE DYADS.

§ 1. Oxygen.

Symbol O. *Atomic weight* 16. *Equivalence* II. *Density* 16. *Molecular weight* 32. *Molecular volume* 2. *One liter weighs* 1·43 *grams (* 16 *criths).*

160. History.—Oxygen was discovered by Priestley in 1774, who called it dephlogisticated air. The following year Scheele discovered it independently, and gave it the name empyreal air. Condorcet called it vital air. After the overthrow of the theory of phlogiston by Lavoisier in 1781, he gave it the name oxygen, from the Greek ὀξύς and γεννάω, acid-former.

161. Occurrence.—Oxygen is the most abundant element in nature. It exists free in the atmosphere, of which it forms a fifth part. Combined with other elements, it constitutes two-thirds of the entire globe. Water is eight-ninths oxygen, silica is one-half oxygen, and alumina one-third oxygen, by weight. Fully one-half of the weight of all minerals, three-quarters of the weight of all animals, and four-fifths of the weight of all vegetables, is oxygen.

162. Preparation.—The atoms of compound molecules containing oxygen may be re-arranged so as to yield simple oxygen molecules :—

I. By the action of some physical force; as

(a) Heat.—Mercuric oxide when exposed to a high temperature, is resolved into mercury and oxygen :—

$$\begin{matrix} Hg=O \\ Hg=O \end{matrix} \quad \text{becomes} \quad \begin{matrix} Hg \\ \| \\ Hg \end{matrix} \quad \text{and} \quad \begin{matrix} O \\ \| \\ O. \end{matrix}$$

(b) Light.—Carbonic di-oxide in the leaves of plants, yields under the influence of sunlight, carbon and oxygen :—

$$(CO_2)_2 = C_2 + (O_2)_2.$$

(c) Electricity.—In the electrolysis of oxygen compounds.

II. By the chemism of some other element, assisted by some physical force, generally heat: as when chlorine acts upon the vapor of water at a red heat, producing hydrogen chloride and oxygen :—

$$(H_2O)_2 + (Cl_2)_2 = (HCl)_4 + O_2,$$

or upon calcium oxide or lime, producing calcium chloride and oxygen :—

$$(CaO)_2 + (Cl_2)_2 = (CaCl_2)_2 + O_2.$$

Experiments.—The original experiment by which Priestley discovered oxygen, is an interesting one, and may be performed with the apparatus shown in Fig. 15. An ordinary test-tube—coated with copper by electro-deposition for about two inches from the sealed end—is used to contain the mercuric oxide, and is supported above the gas-burner by a clamp. By means of a cork and tube, the oxygen as evolved, is conducted beneath the mouth of a glass receiver filled with water, standing in the porcelain cistern.

The usual method of obtaining oxygen is by heating potassium chlorate, when the following reaction takes place :—

$$(KClO_3)_2 = (KCl)_2 + (O_2)_3.$$

But, since this decomposition takes place at a very elevated temperature, and then with almost explosive rapidity, it is found convenient in practice to mix the salt with one-fourth of its weight of some metallic oxide, such as ferric or cupric oxide, or manganese di-oxide. The mode in which these oxides act is at present obscure; but by their use the oxygen is evolved with great uniformity and at

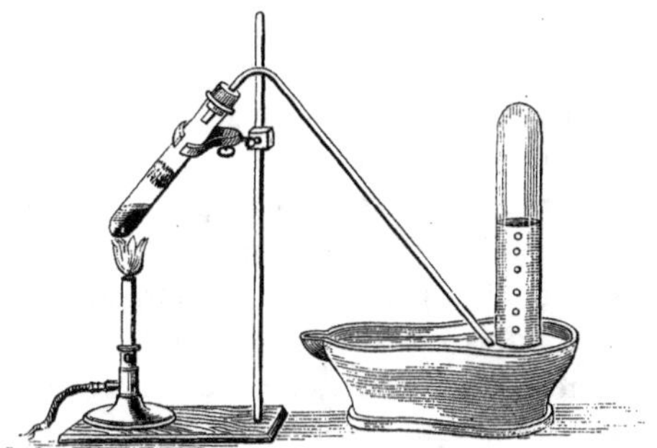

Fig. 15. Preparation of Oxygen from Mercuric oxide.

a far lower temperature, though it is not quite as pure. The oxide is apparently unchanged by the operation. The apparatus employed is shown in Fig. 16. The mixture of potassium chlorate and manganese di-oxide is placed in a flask standing in a sand-bath, through the cork of which a bent glass tube passes to convey the

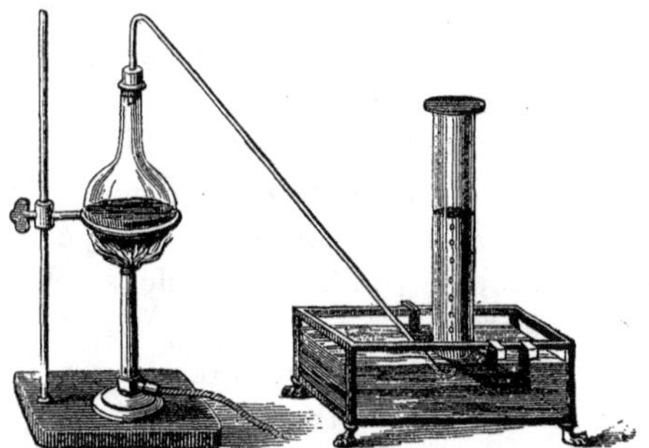

Fig. 16. Preparation of Oxygen from Potassium chlorate.

oxygen as set free, to the cylinder previously filled with water and standing in the water-cistern. Upon applying heat the gas is regularly evolved and may be collected for examination over water as shown in the figure, or, as it is heavier than air, by displacement. When larger quantities of the gas are required, the glass flask may be advantageously replaced by a conical flask with a flat bottom, made of sheet iron.

163. Properties.—I. PHYSICAL.—Oxygen is a colorless, odorless, and tasteless gas. It is somewhat heavier than air, its specific gravity being 1·1056. Water dissolves it slightly, 100 volumes taking up 3 of oxygen at ordinary temperatures. It is incoercible, never having been liquefied by cold or pressure. Its refractive power is to that of air as 0·8616 to 1. It is strongly magnetic; calling the magnetism of iron 1,000,000, that of oxygen is 377. Hence the magnetic power of the atmospheric oxygen is quite appreciable, being equal to that of a layer of iron covering the earth to the thickness of 0·1 millimeter. The diurnal variations of the magnetic needle are due, at least partially, to variations in the intensity of this magnetism, owing to changes of temperature.

II. CHEMICAL.—Oxygen is capable of entering into combination with all the elements but fluorine. But, in the state in which it is usually obtained, heat is necessary to bring about this union. Combustion, in the ordinary use of that term, is union with oxygen, attended with light and heat. When hydrogen, sulphur, charcoal, phosphorus, sodium, and iron, for example, are brought in contact with oxygen at a suitable temperature, they burn, evolving heat and light, and producing oxides of these substances. Oxygen is therefore an intensely active substance, in which the rapidity of ordinary combustion is vastly increased. It is respirable when pure, and produces a quickening of the circulation.

EXPERIMENTS.—A lighted candle burns far more brilliantly in oxygen than in air. If it be blown out, leaving a spark upon the wick, it is immediately rekindled in oxygen gas, with a slight puff. It is by this means indeed, that this gas is recognized; a sliver of wood with a spark upon the end, bursting into flame in oxygen. In this way, a jar, filling with the gas by displacement, may be from time to time tested.

A piece of charcoal—that from oak or spruce is best—having a spark upon it, bursts into vivid combustion when placed in a jar of oxygen. Sulphur, lighted and introduced in a combustion-spoon, burns with a bright blue flame. Sodium heated to redness, burns with a dazzling light. Iron—used in the form of watch spring or of small wire, to which the end of a match is tied—burns in oxygen with great activity.

The most brilliant experiment with oxygen is the combustion in it of phosphorus. A very neat apparatus for this purpose is shown in Fig. 17. A light wire tripod has a ring at its upper part for supporting a globe to contain the gas, and just below it a shallow cup, containing water, into which the neck of the globe enters. From the center of this cup, rises a wire crowned by a small hemispherical cup to receive the phosphorus. The globe is filled about four-fifths with oxygen by displacement, and then inverted on the stand. When all is ready, a piece of phosphorus is cut from a solid stick, and *very thoroughly dried* between sheets of blotting or filtering paper. The globe is raised, the piece of phosphorus dropped into the cup, inflamed by a hot wire, and the globe replaced. The combustion is at once exceedingly vivid; but, in a few seconds, the phosphorus becomes volatilized by the heat, and then burns throughout the entire mass of the oxygen, with a brilliance almost inconceivable.

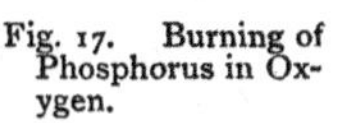

Fig. 17. Burning of Phosphorus in Oxygen.

If now, after cooling, water be added to the jars in which these combustions have taken place, a direct union will take place between the water and the oxide produced. Carbon produces carbonic di-oxide; sulphur, sulphurous oxide; sodium, sodium oxide; iron, tri-ferric tetr-oxide; and phosphorus, phosphoric oxide. With water, the negative oxides of carbon, sulphur, and phosphorus, yield carbonic, sulphurous, and phosphoric acids; and on testing the water with a solution of blue litmus, it will be found to be reddened. With water the positive oxide of sodium yields sodium base, and this turns a solution of red litmus blue. The oxide of iron produced does not form hydrates, and hence in this jar the litmus is unaffected.

164. Uses.—Oxygen is used in the arts for increasing the intensity of combustion, for purposes either of heat or light. Various methods have been proposed for its manufacture from the air on the large scale, the best of which are: (1) Tessie du Motay's, which consists first, in passing pure air over a heated mixture of manganese di-oxide and sodium hydrate, producing by its oxygen sodium manganate; and second, in heating this manganate still higher, by which, with the aid of a current of steam, it is decomposed into the original materials again, setting the oxygen free. (2) Mallet's, in which cuprous chloride is oxidized by passing air over it at a high temperature, to cupryl chloride, which at a higher temperature becomes cuprous chloride and oxygen again. And (3) Deville's, which consists in the decomposition of sulphuric acid by heat.

In the natural world, the uses of oxygen are well nigh infinite. Diluted with nitrogen in the air, it is continually entering and leaving chemical combinations, setting free in the former and absorbing in the latter, enormous stores of force. It is by the oxygen of respiration that the force of living beings is set free from their food; it is by the separation again of oxygen by the sunlight, that that force is stored up anew in this food. Faraday has calculated that 6,000,000,000 pounds of oxygen are daily consumed in the respiration of animals; and that the daily consumption of oxygen for all purposes whatever, reaches the enormous sum of 7,142,857 tons! We have not however to fear an exhaustion of the supply, since the air of the globe, were no oxygen added to it, contains enough of this gas to supply this enormous demand for 480,000 years.

OZONE.—*Molecular formula* O_3. *Molecular weight* 48. *Molecular volume* 2. *Density* 24. *One liter weighs* 2·15 *grams* (24 *criths*).

165. History.—Oxygen, like chlorine, is capable of existing in both passive and active states. The passive condition is the one we have just considered. The active conditions are two in number, termed ozone and antozone. As early as 1785, von Marum noticed that oxygen, on being electrified, acquired an odor strongly resembling that perceived after a stroke of lightning, and usually termed "sulphurous." It was not until 1840, however, that any accurate experiments were made upon the subject. Then Schönbein noticed the similarity between the electrical odor and that produced in the electrolysis of water, and in the slow oxidation of phosphorus and of sulphur; and showed that in each of these cases, the substance produced turned paper moistened with a solution of potassium iodide and starch, to a deep blue. The same year Marignac and De la Rive proved this substance to be modified oxygen. In 1852, Becquerel and Fremy showed that pure oxygen could be entirely converted into ozone. In 1860, Andrews and Tait showed that a contraction of volume took place when oxygen became ozone; and in the same year Soret showed that oil of turpentine absorbed the entire ozone molecule, and in this way determined its density; confirming his results in 1867 by the method of diffusion of gases. The same year Andrews proved that the substance in the air which affected test-papers, was ozone.

The existence of antozone was proposed as a hypothesis by Schönbein in 1858; its actual existence and properties were ascertained by Meissner in 1863, and more fully investigated by him in 1869.

166. Preparation. — Oxygen may be condensed to ozone :—

I. By physical methods; as

(a.) Heat; as when a spiral of platinum wire is heated in air.

(b.) Light; as when essential oils become strongly ozonized in the sunlight. Or, as when the oxygen set free from growing plants by sunlight, contains ozone.

(c.) Electricity; 1st, by the electric silent discharge through oxygen; 2d, in the electrolysis of water acidulated with sulphuric and chromic acids.

II. By the chemical process of slow combustion, and in some cases, of active combustion also. And by the decomposition of barium peroxide and potassium permanganate by sulphuric acid.

EXPERIMENTS.—For obtaining ozone by the action of the silent electric discharge, the apparatus of Siemens, Fig. 18, is the most satisfactory. As shown in the diagram above the cut, it consists of an inner and an outer tube of glass, the inner surface of the inner tube and the outer surface of the outer tube being covered with tin-foil. Between the two tubes a slow current of pure dry oxygen passes, while the two metal surfaces are connected with an active Ruhmkorff coil. In this way 15 per cent of the oxygen may be converted into ozone, a far larger amount than by any other method.

A small flask half full of oil of turpentine, exposed freely to the air and full sunlight for many weeks, acquires powerful ozonizing properties. Sometimes the ozone is directly given up to other bodies; but often it is not so surrendered, except by the intervention

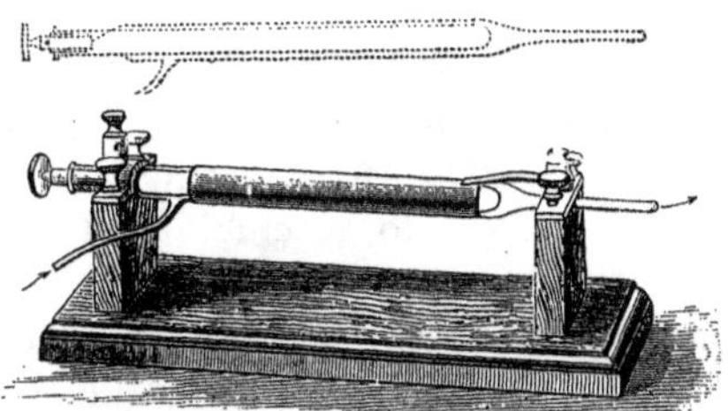

Fig. 18. Siemens's Tube for Ozonizing Oxygen.

of a third body, called therefore, an ozone-carrier. Platinum sponge and ferrous salts act as such carriers; and blood corpuscles are specially active.

To produce ozone by the slow oxidation of phosphorus, place in a perfectly clean and spacious gas-jar, a piece of phosphorus a centimeter in diameter and three or four long, previously scraped clean, and pour in water enough to half cover it. The jar is then loosely stoppered and left to itself, at the ordinary temperature, for several hours. The air in the jar will then be found strongly ozonized.

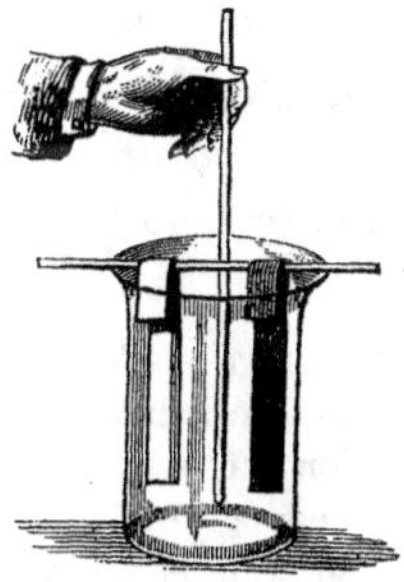

To produce ozone by the slow combustion of ether, a few drops of ether are poured into a beaker, as shown in Fig. 19, across top of which is placed a rod on which hang two slips of test-paper, one of blue litmus, for acids, the other for ozone. If now a glass rod, previously heated to a high temperature, be thrust into the jar, the ether will undergo slow combustion, generating acid-vapors which redden the litmus paper, while the ozone which is formed at the same time, will turn the other paper blue.

Fig. 19. Ozone by Slow Combustion of Ether.

If a vigorous blast of air be directed from a glass tube through the extreme top of the flame of a Bunsen gas-burner, into a capacious beaker, the air in the beaker will give the reaction for ozone; an evidence that it is produced in rapid combustion.

167. Properties.—Physically, ozone is like oxygen, except in density; it is half as heavy again as that gas. Chemically too, it is like oxygen, in the fact that all its compounds are oxides. In a word it is oxygen with properties intensified; it is active oxygen. It has a strong odor, which is said to resemble that of weak chlorine; hence its name, from ὄζω, to smell. At a temperature of 290° it is re-converted into ordinary oxygen. Its most remarkable property is its powerful oxidizing action. It bleaches strongly, carries silver up to the peroxide, and is very poisonous to animal life, on account of its irritating action upon the mucous surfaces. It is soluble only in oil of turpentine. It decomposes potassium iodide, oxidizing the potassium and setting the iodine free.

168. Tests.—Schönbein's test consists of paper moistened with a dilute solution of potassium iodide and starch. The iodine is set free by the ozone, and colors the starch deep blue. Fremy's test is paper moistened with an alcoholic tincture of guaiacum; it is turned light-blue by ozone. Paper moistened with manganous sulphate, or lead hydrate, becomes dark brown or black in ozone.

EXPERIMENTS.—To prepare Schönbein's test-paper, one part of pure potassium iodide is dissolved in two-hundred parts of water, the solution gently heated, ten parts of fine starch gradually added, and the heating continued until the whole becomes homogeneous. Slips of filtering paper are then drawn through the solution and dried in the air. They must be kept in a closely stoppered bottle. When one of these slips is moistened and exposed to an atmosphere containing ozone, it becomes deep blue.

The bleaching action of ozone may be shown by agitating in a jar of air ozonized by phosphorus, a little very dilute solution of indigo. It is at once decolorized.

169. Occurrence and Uses.—Ozone is found free in the air, especially after a thunderstorm. It is also produced by decay, and probably by plant growth. It acts to oxidize and destroy impurities in the air. One volume of air containing $\frac{1}{6000}$ of ozone will purify 540 volumes of putrid air. Atmospheric ozone burns up miasmatic exhalations and hence preserves the air pure.

In the arts it has been used as a disinfectant, and also as a bleaching agent.

Antozone is produced simultaneously with ozone, by the action of the silent electric discharge upon oxygen. On passing the electrized oxygen through a solution of potassium iodide, the ozone is absorbed, and the antozone, mixed with the excess of unaltered oxygen, remains. This, when passed through water, unites with it to form a peculiar, dense mist, which, when collected in a separate vessel and left to itself, disappears in the course of an hour depositing only pure water. Antozone reverts to common oxygen by the same agencies which cause this result with ozone. In presence of ozone, however, the conversion is more rapid.

It is probable that, while the molecule of oxygen is diatomic, that of ozone is triatomic and that of antozone, monatomic, thus:—

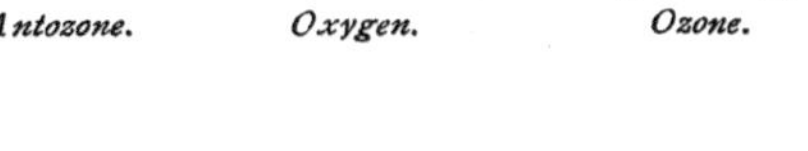

O>	O=O	O / \ O — O
Antozone.	*Oxygen.*	*Ozone.*

COMPOUNDS OF OXYGEN WITH HYDROGEN.

HYDROGEN OXIDE.—*Formula* H_2O. *Molecular weight* 18. *Molecular volume* 2. *Density* 9. 1 *liter weighs* 0·8064 *grams*, (9 *criths*). *Specific gravity* 1. *Solidifies at* 0°. *Boils at* 100°.

170. History.—Hydrogen oxide, or water, was considered to be an elementary substance until 1773, when Lavoisier showed its compound nature. Cavendish and Watt in 1781, first proved its composition by synthesis. In 1805, Humboldt and Gay-Lussac ascertained that the ratio of its constituents by volume, is as 2 : 1; and Berzelius and Dulong proved that the ratio by weight is as 1 : 8.

171. Occurrence.—Water occurs abundantly diffused in nature, both free and in combination. Natural waters are seldom pure; even the water which falls as rain contains atmospheric impurities. It is essential to the life of plants and animals, and enters into the composition of many mineral substances. Seven-eighths of the entire human body is water.

172. Preparation.—Hydrogen oxide may be prepared synthetically; that is by the direct union of its constituent elements. The product of the combustion of hydrogen is always water, as we have seen. (Fig. 6.) And when the two gases are mixed together in the ratio of two volumes of hydrogen to one of oxygen, they may be caused to unite by a flame, by an electric spark, or by finely divided platinum. The heat evolved by their union is very great; when the two gases are burned together in the above proportions from a jet, they give the most intense heat which can be obtained by combustion. This experiment was first made by Hare of Philadelphia, in 1801; and the apparatus is called the compound or oxy-hydrogen blowpipe.

EXPERIMENTS.—To show the production of water by the combustion of hydrogen, the experiment described under hydrogen may be repeated. In Fourcroy's experiment, the gas continued to burn for a week, consuming 37,500 cubic inches of oxygen and hydrogen and producing 15 ounces of pure water.

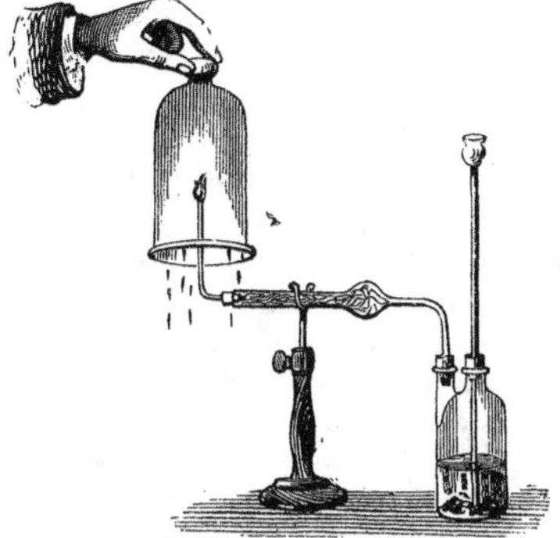

Fig. 20. Water from the combustion of Hydrogen.

To show the union of the mixed gases by flame, a mass of soap-bubbles may be blown in a metallic dish containing soap and water, by a bubble-pipe attached to a gas-bag containing one volume of oxygen to two of hydrogen. On applying a flame (Fig. 21), the gases explode with a loud report.

For the purpose of measuring exactly the proportions in which the gases unite, an instrument called a eudiometer is employed.

Fig. 21. Explosion of mixed Oxygen and Hydrogen gases.

Fig. 22 represents the form proposed by *Ure;* it is simply a U-shaped tube of glass which is closed at one end, the closed limb being graduated, and pierced near its extremity by two platinum wires. This limb is to be filled with water, and then a given quantity of pure oxygen, say 20 cubic centimeters is to be introduced from a delivery-tube;

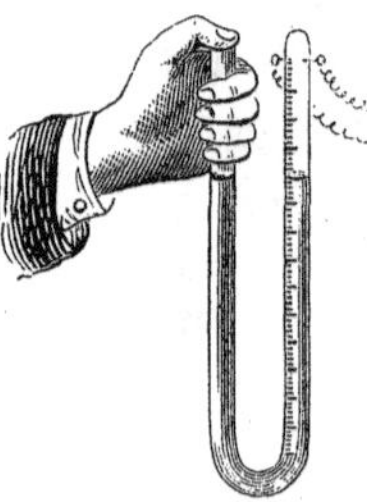

Fig. 22. Ure's Eudiometer.

50 cubic centimeters of hydrogen are then similarly introduced,—all measurements being made when the level of the liquid is the same in both limbs—the open end is closed firmly by the thumb, as shown in the figure, —a cushion of air being left between it and the liquid,—and a spark passed through the mixed gases by means of the platinum wires. Upon restoring the level of the liquid by adding water, 10 cubic centimeters of gas will be left, which, on examination, will be found to be pure hydrogen. Hence 20 volumes of oxygen have united with 40 of hydrogen to form water. But this is the ratio of 1 : 2, which theory requires.

The remarkable action of spongy platinum in causing the union of oxygen and hydrogen gases, may be shown by holding a piece of this metal—or what answers equally well, a piece of asbestus previously moistened with strong platinic chloride solution, and heated to redness—over a jet from which hydrogen is issuing. The mass becomes at once red-hot and fires the gas. This effect is attributed to the enormous surface-attraction which platinum, in this form, has for gaseous substances. Exposed to the air, platinum-sponge condenses oxygen in this way within it, perhaps to the state of a liquid. When now, it is exposed to hydrogen, it condenses this gas also; thus bringing them together and causing their union.

The great heat evolved by burning oxygen and hydrogen gases together, requires for its production a concentric jet, consisting of an inner tube carrying the oxygen, and outside of this, a second tube between which and the first the hydrogen passes. The two gases must be brought from separate gas-holders. The hydrogen is first lighted; it burns with a large yet pale flame. On admitting the oxygen, this flame becomes smaller, and is drawn out very long and fine, being altered also in color. If this flame be directed upon various metals contained in small cups of charcoal, they may be melted and burned, each with its characteristic color. The brilliance of the experiment is much heightened in many cases, by shutting off the hydrogen and allowing the combustion to take place in the oxygen alone. A watch-spring or a small file, introduced into the flame, burns with vivid scintillations. A piece of cast-iron on charcoal, gives, after melting it and shutting off the hydrogen, a superb pyrotechnical effect. On introducing some infusible substance as a pipe-stem, a cylinder of magnesia or zirconia, or still better, one of lime, the light emitted is dazzling. This light was first utilized practically in the trigonometrical survey of Great Britain, by Lieut. Drummond, when it was seen 108 miles in full daylight. It is sometimes called the Drummond light; but is more properly called the calcium, or oxy-hydrogen light. By means of the jet of flame now described 3200 ounces of platinum have been melted in one operation.

But not only may the composition of water be established by synthesis; it may be equally well determined by analysis. For this purpose both direct and indirect means may be employed.

EXPERIMENTS.—In the sodium experiment (Fig. 1) the hydrogen set free must have been derived from the water on which the sodium acted. If a little solution of red litmus be added to the water after the experiment, it will be blued. A base must therefore have been produced by the sodium; but a base contains oxygen, which oxygen must also have come from the water. In this way the composition of water may be established by an indirect analysis.

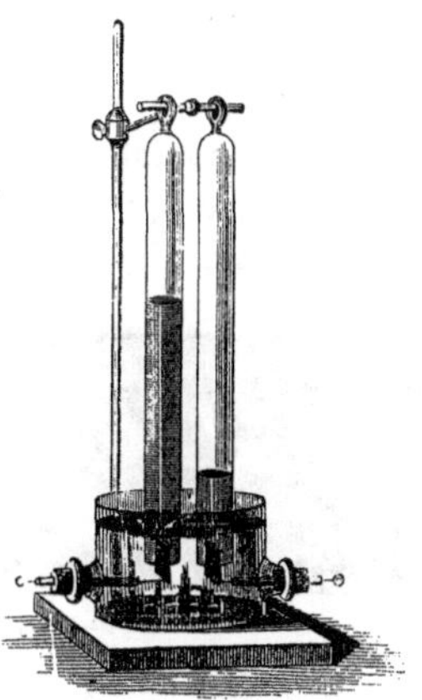

Fig. 23. Decomposition of Water by Electricity.

To analyze water directly, it may be submitted to the action of electricity. But as water itself is not decomposed by this agent, a secondary action must be made use of. A little sulphuric acid is added to the water; this is decomposed by the electricity, and reacts upon the water, separating it into its constituents. A convenient apparatus for this purpose is shown in Fig. 23. Two tubes closed at one end and filled with water, are suspended with their mouths beneath the surface of some acidulated water contained in the glass dish below. Through the sides of this dish two wires pass, each terminating in a plate of platinum seen beneath the open ends of the tubes. On connecting these wires with a Bunsen's or Grove's battery of 6 or 8 cells, a torrent of gas-bubbles rises from each platinum plate into the tube placed above it. It will soon be noticed that the tube over the negative electrode fills twice as rapidly as the other; and on testing the gas in each, when the tubes are both full, the gas in this tube will be found to be hydrogen, while that in the other is oxygen. Water contains therefore 2 volumes of hydrogen and 1 volume of oxygen. And as oxygen is 16 times heavier than hydrogen, the ratios by weight must be as 2 : 16 or as 1 : 8; or more exactly, water consists of 88·89 per cent of oxygen and 11·11 per cent of hydrogen.

173. Properties.—Hydrogen oxide is a limpid, colorless liquid, without odor or taste. It is neither acid nor alkaline in its action on vegetable colors, and is a poor conductor both of heat and electricity. It is 773 times heavier than air at 0°. It is the standard of specific gravity for liquids, and is taken as the unit of weight in the decimal system, one cubic centimeter of water at 4° weighing 1 gram; hence 1 liter of water at the same temperature weighs 1 kilogram. When cooled to 0°, it solidifies in crystals which are derived from the hexagonal prism, and, which are often very beautifully seen in snow-flakes, Fig. 24. When heated to 100° it is con-

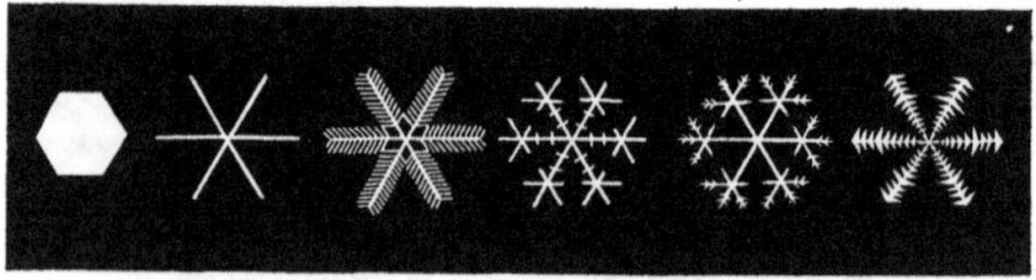

Fig. 24. Snow Crystals.

verted into vapor called steam. The rate of its expansion increases slightly with the temperature; though at

4°, water reaches its point of maximum density, and then if cooled below this, it expands until its solidifying point is reached. At the moment of becoming solid it increases considerably in volume, 916 cubic centimeters of water becoming 1,000 of ice. Its index of refraction at 0° is 1·333. It is also the standard of specific heat, since it requires more heat to raise its temperature a given number of degrees than any other solid or liquid. In the form of steam it is a colorless gas, having a density of 9, or a specific gravity of 0·622. One volume of water yields 1700 volumes of steam at 100°.

Chemically, water is a very active substance. It enters into combination directly with most positive and negative oxides, forming bases with the former and acids with the latter, evolving more or less heat. A familiar example of this is the slaking of lime, a process which may be represented by the following equation :—

$$\underset{\text{Calcium oxide.}}{CaO} + \underset{\text{Water.}}{H_2O} = \underset{\text{Calcium hydrate.}}{Ca''(OH)_2}.$$

Water also enters molecularly into the composition of many crystalline substances, the amount appearing to increase in proportion as the crystallization takes place in a colder and more dilute solution. Calcium sulphate crystallized takes two molecules, $CaSO_4$, 2 aq.; copper sulphate, five, $CuSO_4$, 5 aq.; magnesium sulphate, seven, $MgSO_4$, 7 aq.; sodium sulphate, ten, Na_2SO_4, 10 aq.; and potassio-aluminic sulphate (alum), twenty-four, $K_2Al_2^{VI}(SO_4)_4$, 24 aq. Such crystals, when exposed to dry air, *effloresce*; i. e., lose this water of crystallization and fall into a white powder. On the other hand, some substances, in a moist atmosphere, attract water and liquefy: this is called deliquescence. The solvent power of water is very much greater than that of any other liquid. Each substance which it dissolves however, has a fixed limit of solubility, which depends upon temperature, etc. Gaseous solubility is to a very large extent dependent upon atmospheric pressure also.

174. Natural Waters.—The purest natural water which can be obtained is that which falls as rain; but even this is contaminated with matters washed from the air. Other natural waters may be divided into potable (or drinkable), mineral, and saline waters. Of potable waters, river and lake waters, especially such as are found in granitic regions, are the purest. That of Loch Katrine in Scotland, containing but 2 grains of solid matter to the gallon, is one of the purest waters known; while the purest water supplied to any city in this country is that from Lake Cochituate which supplies Boston, which contains but 3·11 grains in one gallon. The Schuylkill water (Philadelphia) contains 3·50 grains; Ridgewood (Brooklyn) 3·92; the Croton (New York) 4·78; Lake Michigan (Chicago) 6·68; the water which supplies Albany, 10·78; and that of the Thames, which supplies London in part, 16·38 grains, in each gallon. Spring and well waters are seldom as pure as surface waters, since they have penetrated the ground and taken up solid impurities. Thus the water of a well near Central Park, New York, gave 43·54 grains; one in Schenectady, 49·21 grains; one in Amsterdam, 68·93 grains; and one in London, 99·97 grains of solid matter to the gallon. Mineral waters are classified, according to their prevailing constituents, into sulphurous, chalybeate, alkaline, etc. The amount of solid matters which they hold in solution varies very widely; the chalybeate spring of Tunbridge Wells contains but 7 grains to the gallon, while the Saratoga Seltzer spring contains over 400, one of the springs at Vichy, 460, the High Rock spring at Saratoga, 628, and the artesian Lithia spring at Ballston, 1233. Saline waters, especially those of inland lakes with no outlet, are most impure. Sea-water contains on an average, 2,500 grains of solid matter, the water of the Dead Sea 12,600 grains, and that of the Great Salt Lake 22,000 grains to the gallon.

HYDROGEN PEROXIDE. — *Free hydroxyl. Formula* H_2O_2. *Graphic* H—O—O—H. *Molecular weight* 34. *Specific gravity of liquid,* 1·452.

175. History.—Hydrogen peroxide was discovered by Thenard in 1818, and called by him oxygenated water.

176. Preparation and Properties.—It is always prepared from barium peroxide by the action of hydrochloric, or better of carbonic acid. The reaction is:—

$$BaO_2 + H_2CO_3 = BaCO_3 + H_2O_2.$$

By adding the materials alternately, the water present soon becomes saturated; then, by evaporation over sulphuric acid, this water may be removed, thus leaving pure hydrogen peroxide.

It is always formed by slow oxidation in presence of moisture; the antozone thus produced uniting with the water to give hydrogen peroxide, which may be detected in the liquid.

Hydrogen peroxide is a colorless syrupy liquid, of specific gravity 1·452. It does not solidify at—30°, and may be evaporated in vacuo unchanged. It begins to decompose at 15°, and at 100° separates into water and oxygen with almost explosive violence. It is more permanent if diluted with water. It has a harsh taste, and whitens the skin when placed upon it. It bleaches vegetable colors.

Its most remarkable property is the facility with which it evolves oxygen under certain conditions. Metallic platinum, gold, and silver, when finely divided, decompose it almost with explosion; their oxides, as well as the peroxides of lead and manganese, also decompose it, giving up a part of their oxygen at the same time. It is therefore at once an oxidizing and a reducing agent.

EXPERIMENTS.—The water surrounding the phosphorus in the preparation of ozone (p. 123) contains hydrogen peroxide and will reduce a dilute solution of potassium permanganate, itself a strong oxidizing agent. If to a solution containing hydrogen peroxide, a few drops of a dilute solution of potassium chromate be added, and the whole be agitated with ether, the etherial layer which forms above the liquid on standing, will be blue from the presence of perchromic acid.

OXIDES AND ACIDS OF CHLORINE.

This element, as already mentioned, may act as a monad, a triad, a pentad, or a heptad. Its oxygen compounds, together with their corresponding hydrates or acids, are therefore, as follows:—

Hypochlorous oxide	Cl'_2O	Hypochlorous acid	$HCl'O$
Chlorous oxide	Cl'''_2O_3	Chlorous acid	$HCl'''O_2$
Chloric oxide	$Cl^{V}_2O_5$	Chloric acid	$HCl^{V}O_3$
Perchloric oxide	$Cl^{VII}_2O_7$	Perchloric acid	$HCl^{VII}O_4$

Of these oxides, only the first two have been prepared. Of the acids, all have been obtained.

176. Hypochlorous Oxide and Acid. — Hypochlorous oxide was discovered by Balard in 1834. It may be prepared by passing dry chlorine gas over mercuric oxide :—

$$HgO + Cl_4 = HgCl_2 + Cl_2O.$$

A yellow gas is obtained, which condenses to a blood-red liquid at $-20°$. This gas has a penetrating, chlorine-like odor, and is decomposed with explosion by very slight causes, yielding two volumes of chlorine and one of oxygen. It is very soluble in water, uniting with it to form hypochlorous acid, $Cl(OH)$, which retains the odor of the oxide and is a powerful bleaching agent, twice as active as chlorine.

Hypochlorites are prepared in the arts by exposing alkaline hydrates to the action of chlorine gas. With sodium hydrate, the reaction which takes place is as follows :—

$$(HNaO)_2 + Cl_2 = NaClO + NaCl + H_2O.$$

By treating a hypochlorite with dilute nitric acid and distilling, hypochlorous acid may be obtained.

Hypochlorites are used very largely in the arts as bleaching agents ; the so-called chloride of lime, or calcium hypochlorite, being manufactured for this purpose on an immense scale.

177. Chlorous Oxide and Acid.—Chlorous oxide was first described by Millon. It is prepared by acting upon a chlorate with nitric acid in presence of a reducing agent, like arsenous acid. It is a yellowish-green gas, having a specific gravity of 2·65, bleaching indigo and litmus, and soluble in one-sixth its volume of water, forming chlorous acid. It is not condensed to a liquid by a cold of $-20°$. It explodes when heated to $57°$ and also when brought in contact with sulphur, phosphorus, or arsenic.

Chlorous acid, $ClO(OH)$, combines slowly with bases, forming chlorites, which are unstable, breaking up easily into chlorates and chlorides.

178. Chlorine Tetr-oxide. — Cl_2O_4 or $O_2Cl^V—Cl^VO_2$. Free chloryl. This oxide, though intermediate between chlorous and chloric oxides, is constituted very differently, its chlorine atoms being directly united. It was discovered by Davy, in 1814, and is obtained by the action of sulphuric acid upon potassium chlorate, at a low temperature. A dark-greenish gas is evolved, which, strongly diluted, has a sweetish aromatic odor, and is strongly oxidizing in its action. At —20° it condenses to an orange-red liquid. It explodes with great violence above 60°, often spontaneously. It has no corresponding acid.

EXPERIMENTS.—The energy of its action on combustibles may be shown by mixing together on a sheet of paper, about a gram of finely pulverized potassium chlorate and an equal quantity of fine sugar. Place the mixture on a fragment of brick, and touch it with a glass rod previously dipped in sulphuric acid. The chlorine tetr-oxide thus set free causes a vivid combustion of the entire mass.

Fig. 25. Burning of Phosphorus by Chlorine tetroxide.

Or, a gram of chlorate in crystals may be placed at the bottom of a conical glass filled with water, Fig. 25, a few small pieces of phosphorus added, and sulphuric acid allowed to come in contact with the salt, by means of a pipette. The phosphorus at once takes fire in the chlorine tetr-oxide gas evolved, and burns vividly.

179. Chloric Acid.—Chloric acid, $ClO_2(OH)$, was first prepared by Gay-Lussac. On passing chlorine through a solution of potassium hydrate, potassium chlorate and potassium chloride are obtained, according to the equation :—

$$(Cl_2)_3 + (HKO)_6 = KClO_3 + (KCl)_5 + (H_2O)_3.$$

By adding to a solution of potassium chlorate, fluosilicic acid, or to one of barium chlorate, sulphuric acid, the potassium or barium is separated, and there is left an aqueous solution of chloric acid which by concentration in vacuo, may be obtained as a colorless, syrupy, acid liquid, which decomposes above 40°, and is a strongly oxidizing agent. Sulphur, phosphorus, alcohol, paper, are at once inflamed by it. Its salts, the chlorates, are also active oxidizing agents. They are used for the preparation of oxygen, and in detonating and pyrotechnical mixtures.

EXPERIMENTS.—Mix carefully on paper half a gram of fine potassium chlorate with quarter of a gram of sulphur. Wrap up the mass in paper, place it on an anvil and strike it with a hammer. It will explode violently. Many new explosives have been recently proposed, consisting of potassium chlorate mixed with tannin, with catechu, and with potassium ferrocyanide and sugar. They are all more or less unstable and therefore dangerous. A mixture of amorphous phosphorus and potassium chlorate sometimes detonates spontaneously.

180. Perchloric Acid.—Perchloric acid, $ClO_3(OH)$, was discovered by Stadion; it has been recently more fully examined by Roscoe. On subjecting potassium chlorate to heat, it becomes pasty at a certain stage of the process, and ceases to evolve oxygen. It is then a mixture of potassium perchlorate and chloride, thus :—

$$(KClO_3)_2 = KClO_4 + KCl + O_2.$$

By crystallization the difficultly-soluble perchlorate is

obtained pure; and by distilling this with sulphuric acid, a colorless fuming liquid condenses in the receiver, having a specific gravity of 1·78. It does not solidify at $-35°$. This acid is a powerful oxidizer; it instantly ignites wood or paper when thrown upon it, and is decomposed with explosion, by charcoal. It is the most stable of the chlorine acids.

OXIDES AND ACIDS OF BROMINE AND IODINE.

181. The analogy of bromine and iodine to chlorine is shown also in the similarity of their oxides and acids to those of that element. Theoretically, the series is the same, though only a few members of it have as yet been obtained. Of the bromine compounds, only the following are known:—

Hypobromous acid	$HBr'O$
Bromic acid	$HBr^{V}O_3$
Perbromic acid	$HBr^{VII}O_4$

and of those of iodine, only those given below have been prepared:—

		Hypoiodous acid	$HI'O$
Iodic oxide	$I^{V}{}_2O_5$	Iodic acid	$HI^{V}O_3$
Periodic oxide	$I^{VII}{}_2O_7$	Periodic acid	$HI^{VII}O_4$

182. Iodic Acid.—Iodic acid is the most important of the bodies given above. It is prepared by the direct action of oxidizing agents upon iodine, or by the simultaneous action of chlorine and iodine upon water:—

$$I_2 + (Cl_2)_5 + (H_2O)_6 = (HIO_3)_2 + (HCl)_{10}$$

Iodic acid is a colorless solid, crystallizing in orthorhombic prisms; at 170° it loses a molecule of water, and becomes iodic oxide.

§ 2. SULPHUR.

Symbol S. *Atomic weight* 32. *Equivalence* II, IV, VI. *Density of vapor* 32. *Molecular weight* 64. *Molecular volume* 2. *Specific gravity of solid* 2·04. 1 *liter of sulphur vapor at a temperature of* 1000° *weighs* 2·86 *grams (* 32 *criths).*

183. History.—Sulphur has been known from the remotest times.

184. Occurrence.—It is found free in many volcanic regions, especially in Italy and Sicily. It occurs also in combination, as a constituent of both binary and ternary compounds. The sulphides of iron, copper, lead, zinc, antimony, arsenic, and mercury, are well-known minerals; as are also the sulphates of calcium, barium, strontium, magnesium, and sodium. It forms an essential part of animal tissues and exists to a considerable extent in those of vegetables. Its compounds cause the peculiar odor of cruciferous and alliaceous plants, such as mustard and garlic.

Sicily and Italy yield annually 80,000 tons of sulphur.

185. Preparation.—The sulphur of commerce is the native material purified. As found, it is mixed with various earthy impurities; and to separate it from these it is subjected to heat in earthen pots, as shown in Fig. 26.

These pots are arranged in the furnace in two rows, and are charged from the top. The sulphur is converted into vapor by the heat, passes through the narrow tubes into a second row of earthen vessels which act as receivers, is there condensed to a liquid, and runs out at the bottom into wooden vessels filled with water, placed below. Richer masses are often heated in heaps with just fuel enough to melt the sulphur, which collects in a depression made at the bottom of the heap. The sulphur thus prepared contains still three or four per cent of impurities; it is still further refined by another distillation in cylinders of iron, as shown in Fig. 27. The crude sulphur is melted in a tank by the waste heat of the fire, and then runs down through a pipe into the retort, where it is converted into vapor. This vapor

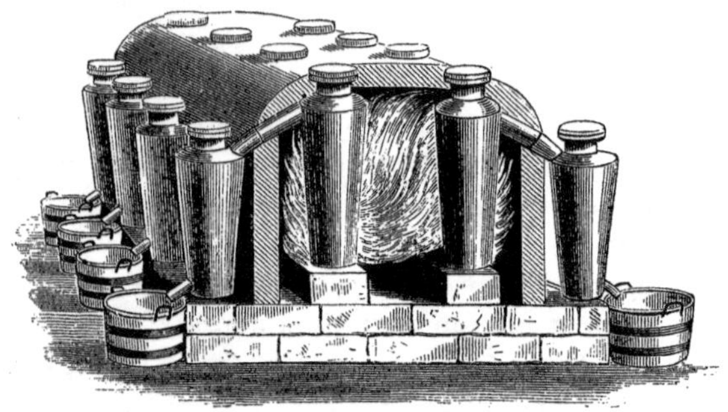

Fig. 26. Crude Distillation of Sulphur.

enters a large brick chamber, and is there condensed. At first, when the walls are cold, a fine powder is produced, known in commerce as flowers of sulphur; but afterwards, when the walls of the chamber become hot, the sulphur condenses to a liquid, which collects on the floor and may be drawn off and ladled into moulds, forming what is ordinarily called roll brimstone.

Sulphur is also obtained from iron disulphide or iron pyrites, which mineral, in some localities, is very abundant. For this purpose the pyrites are piled up in a pyramid with wood, and fire applied. The sulphur which is set free, collects in the liquid form in cavities made in different parts of the heap. Iron pyrites gives up one third of its sulphur when thus treated; yielding about 20 per cent. of its weight.

186. Properties.—I. PHYSICAL.—Sulphur is capable of existing in three distinct allotropic forms or modifications:—

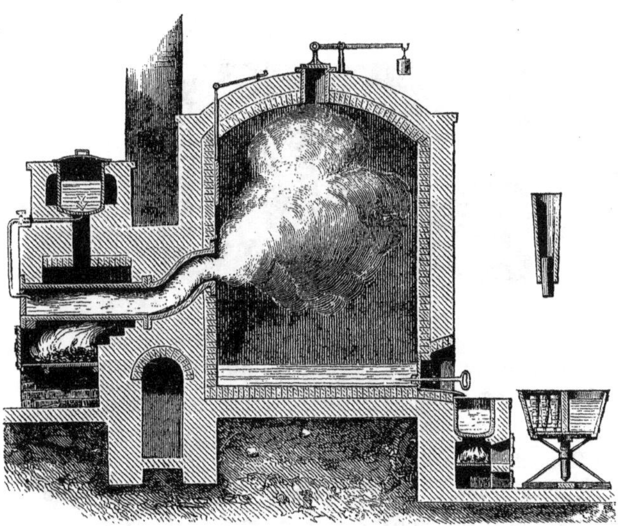

Fig. 27. Refining of Sulphur.

(α.) The first variety is that found in nature. It is a lemon-yellow, brittle solid, crystallizing in orthorhombic octahedrons, often modified (Fig. 28, 1, and 2), and possessing a specific gravity of 2·05. It is easily soluble in carbon disulphide and may be readily crystallized therefrom.

(β.) The second variety is produced by crystallizing

sulphur from fusion, at high temperatures. Yellowish-brown needle-shaped crystals belonging to the monoclinic system, (Fig. 28, 3), are thus obtained, which are transparent and have a specific gravity of 1·98. This variety is insoluble in carbon disulphide, and passes, slowly at ordinary temperatures, more rapidly at higher ones, back into variety *α*.

Fig. 28. Sulphur Crystals.

Sulphur, since it crystallizes in forms belonging to two distinct systems, is called a d i m o r p h o u s element.

(*γ*.) The third variety of sulphur is produced by heating melted sulphur to a temperature of 250° and then suddenly cooling it by pouring it into water. It is a dark brown, tenacious mass, which may be drawn out in threads like caoutchouc. It has a specific gravity of 1·95, and is insoluble in carbon disulphide. It slowly passes into *α* if left to itself; but if heated to 100°, it undergoes this change suddenly, the thermometer rising to 110° from the heat evolved.

Either variety of sulphur melts when heated to 115°, becoming a pale-yellow limpid liquid. As the temperature rises, it becomes viscid, until between 200° and 250° the vessel may be inverted without loss; it then becomes fluid again, and at 440° boils. Its vapor-density was for a long time considered anomalous, being at 500°, 96; but Bineau showed that at 1000° it became normal, 32. The molecule of sulphur at 500° is therefore hex-atomic, while at 1000° it is di-atomic.

EXPERIMENTS.—To produce the β variety of sulphur, melt 250 grams of this substance in a crucible over a gas-flame or in a charcoal fire. Allow it to cool until a crust forms upon the surface. Pierce a hole in this crust near one side, and pour out the sulphur which still remains liquid. The interior of the crucible when cold, will be found lined with needle-shaped crystals, Fig. 29.

Fig. 29. Monoclinic Sulphur Crystals.

Fig. 30. Preparation of Amorphous Sulphur.

The third or amorphous variety of sulphur may be prepared by melting a sufficient quantity in a flask, heating it until the second stage of fluidity is reached, and then pouring it, in a thin stream, into water, as shown in Fig. 30. On removing it from the water, it is found to be remarkably plastic; thus affording an excellent example of allotropism.

II. CHEMICAL.—When heated to 260° in the air, sulphur takes fire, burning with a pale blue flame. It is also a supporter of combustion, many metals taking fire readily in its vapor and burning actively. When united with other elements it forms sulphides.

Berthelot proposes to call the insoluble variety of sulphur electro-positive, and the soluble, electro-negative, because these forms of sulphur are due, in his opinion, to the element with which the sulphur has previously been united. When separated from union with the more negative oxygen, for example, the sulphur is found insoluble and electro-positive; while, obtained from its hydrogen compound, it is soluble and electro-negative.

When combined with positive elements alone, sulphur acts as a dyad and is then the analogue of oxygen. As a dyad too, it may perform a linking function.

187. Tests.—In the free state, sulphur is recognized by its color, by its volatility when heated, and by its odor when burned. In combination, as a soluble sulphide, it blackens paper moistened with a solution of lead.

188. Uses.—Sulphur is employed very largely in the arts for the preparation of sulphuric acid, and for bleaching straws and woolens.

SULPHUR AND HYDROGEN.

HYDROGEN SULPHIDE.—*Formula* H_2S. *Molecular weight* 34. *Molecular volume* 2. *Density* 17. *One liter weighs* 1·52 *grams* (17 *criths*).

189. History and Occurrence.—Hydrogen sulphide—called also hydrosulphuric acid and sometimes sulphurretted hydrogen—was discovered by Scheele in 1777. It occurs in certain volcanic gases, and is the essential constituent of the water of the so-called sulphur springs, such as Sharon and Avon in this country, Harrowgate in England, Bagnieres in France, and Aachen in Germany.

190. Preparation and Properties.—Hydrogen sulphide may be prepared by the direct union of its components, as by passing sulphur-vapor and hydrogen through a red-hot tube, filled with fragments of pumice to increase the heated surface. It is generally obtained by the action of an acid upon some sulphide. When ferrous sulphide, for example, is treated with sulphuric acid at the ordinary temperature, the reaction is :—

$$\underset{\textit{Ferrous sulphide.}}{FeS} + \underset{\textit{Hydrogen sulphate.}}{H_2(SO_4)} = \underset{\textit{Ferrous sulphate.}}{Fe(SO_4)} + \underset{\textit{Hydrogen sulphide.}}{H_2S.}$$

Or, when antimonous sulphide is heated with hydrochloric acid, antimonous chloride and hydrogen sulphide result :—

$$Sb_2S_3 + (HCl)_6 = (SbCl_3)_2 + (H_2S)_3.$$

EXPERIMENTS.—For the continuous preparation of hydrogen sulphide from ferrous sulphide and dilute sulphuric acid, the apparatus shown in Fig. 31 is convenient. It consists of three bulbs of glass, the two lower ones being in a single piece, and the upper one, prolonged by a tube reaching to the bottom of the lower, being ground air-tight into the neck of the second. Through the tubulure of the middle bulb, the ferrous sulphide, in lumps the size of a chestnut, is introduced, the space between the tube and the side of the constriction being too narrow to let them fall through. This tubulure is then closed by a cork through which a glass stop-cock passes. The acid—one part sulphuric acid diluted with fourteen of water—is poured in through the safety tube, runs into the bottom globe, and rises to overflow the iron sulphide in the middle one. If the cock is open, the gas which is evolved escapes ; but when it is shut, the pressure of the accumulating gas forces the liquid from the sulphide, down into the lower, and thence into the

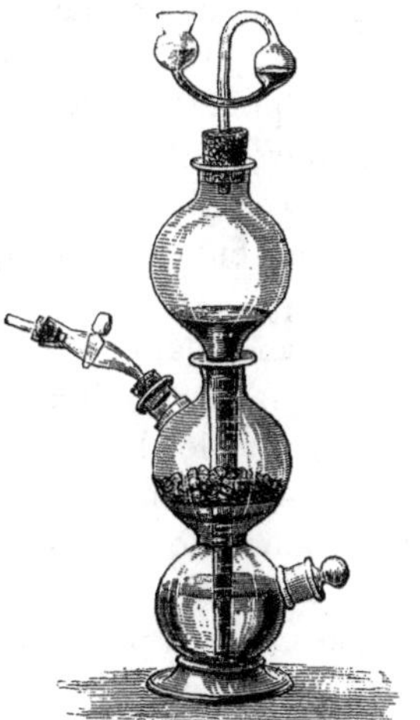

Fig. 31. Hydrogen sulphide Apparatus.

upper bulb, thus stopping the action and preserving a volume of the gas ready for use. By the tubulure of the lower bulb, the acid, when saturated, may be removed.

Hydrogen sulphide is a colorless gas, with a disgusting odor, well known as that of rotten eggs. It is somewhat heavier than air, its specific gravity being 1·177. Cooled to $-74°$, or submitted to a pressure of 17 atmospheres at $10°$, it condenses to a colorless mobile liquid of specific gravity 0·9, which freezes to a mass like ice at $-85°$. It is quite soluble in water, 1 volume of which dissolves 3 volumes at ordinary temperatures, and 4·37 volumes at $0°$. Chemically, hydrosulphuric acid gas is combustible, burning with a pale blue flame. Its reaction with blue litmus paper, is weakly acid. It is easily decomposed by oxidizing agents, the sulphur frequently being deposited. It reacts with metals and their oxides to produce sulphides, setting hydrogen free in the first case, and water in the second. This gas is exceedingly poisonous; according to Faraday, birds die in air which contains but $\frac{1}{1500}$ of it, and dogs in that which contains but $\frac{1}{800}$.

191. Tests and Uses. — Hydrogen sulphide is easily detected by placing in it a strip of paper moistened with a solution of lead acetate; in this way it may be shown to exist in most specimens of coal-gas, and in the gaseous exhalations from drains, cess-pools, and the like. With sodium nitroferrocyanide in alkaline solution, it strikes a deep purple color.

It is used extensively in the laboratory as a re-agent, the sulphides which it produces being characteristic for certain metals, either in color, in solubility, or in some easily recognized property.

Experiment.—The action of hydrogen sulphide upon metallic

solutions may be very well shown by the apparatus represented in Fig. 32. The gas is evolved from ferrous sulphide in the two-necked bottle, and passes successively through the four solutions in the adjoining bottles, the escaping gas being retained in a solution of ammonia. In the first bottle may be placed a dilute acid solution of lead, in the second one of arsenic, in the third one of

Fig. 32. Precipitation of Metals by Hydrogen sulphide.

antimony, and in the fourth one of zinc, the last being made slightly alkaline with ammonia. The sulphide of lead in the first bottle will be black, the sulphide of arsenic in the second, yellow, the sulphide of antimony in the third, orange, and the sulphide of zinc in the fourth, white. The first three metals are precipitated in acid solutions, the last only when the solution is alkaline.

The gas may readily be inflamed by applying to it a lighted taper. By holding a bell-glass over an ignited jet of this gas, it is bedewed with moisture, thus proving that the gas contains hydrogen.

OXIDES AND ACIDS OF SULPHUR.

Sulphur may unite with oxygen as a dyad, tetrad, or hexad, and may therefore form the following series of oxides and acids :—

Hyposulphurous oxide	$S''O$.	Hyposulphurous acid	$H_2S''O_2$.
Sulphurous oxide	$S^{IV}O_2$.	Sulphurous acid	$H_2S^{IV}O_3$.
Sulphuric oxide	$S^{VI}O_3$.	Sulphuric acid	$H_2S^{VI}O_4$.

Hyposulphurous oxide and acid are unknown.

SULPHUROUS OXIDE.—*Formula* SO_2. *Molecular weight* 64. *Molecular volume* 2. *Density* 32. *One liter weighs* 2·86 *grams (*32 *criths).*

192. History and Occurrence.—Sulphurous oxide was first pointed out as a peculiar substance by the alchemist Stahl; but not until 1774, did Priestley carefully examine its properties. It is found among the gaseous products of volcanic action.

193. Preparation.—Sulphurous oxide is uniformly the product of the combustion of sulphur in air or in pure oxygen; being formed thus:—

$$S_2 + O_4 = (SO_2)_2.$$

It is also the product of the action of certain metals such as copper and mercury, upon sulphuric acid. The metal simply displaces the radical of the acid:—

$$\underset{\textit{Copper.}}{Cu} + \underset{\textit{Sulphuric acid.}}{\left.\begin{matrix}(SO_2)\\H_2\end{matrix}\right\}O_2} = \underset{\textit{Copper hydrate.}}{\left.\begin{matrix}Cu''\\H_2\end{matrix}\right\}O_2} + \underset{\textit{Sulphurous oxide.}}{SO_2}.$$

The copper hydrate then reacts with another molecule of sulphuric acid thus:—

$$\underset{\textit{Copper hydrate.}}{\left.\begin{matrix}Cu''\\H_2\end{matrix}\right\}O_2} + \underset{\textit{Sulphuric acid.}}{\left.\begin{matrix}(SO_2)\\H_2\end{matrix}\right\}O_2} = \underset{\textit{Copper sulphate.}}{\left.\begin{matrix}(SO_2)\\Cu''\end{matrix}\right\}O_2} + \underset{\textit{Water.}}{(\left.\begin{matrix}H\\H\end{matrix}\right\}O)_2}.$$

194. Properties.—Sulphurous oxide is a colorless gas, with a pungent suffocating odor, known as that of a burning match. It is more than twice as heavy as air, having a specific gravity of 2·247. Cooled to $-10°$ by a freezing mixture, it condenses to a thin colorless liquid of specific gravity 1·49, which becomes solid at $-76°$. Sulphurous oxide gas is freely soluble in water, one volume of which dissolves at 0°, 68·86, and at 20°, 36·22 volumes of this gas, forming sulphurous acid. This solution when cooled to 0°, deposits cubical crystals consisting of H_2SO_3, 14 aq.

Chemically, sulphurous oxide is neither combustible nor a supporter of combustion; burning bodies introduced into it are at once extinguished. It unites directly with chlorine to form sulphuryl chloride and with positive oxides to form sulphites. Hydrogen sulphite or sulphurous acid has strong acid properties and destroys vegetable colors, apparently by forming a direct compound with them. It exhibits a decided tendency to take up oxygen and to pass into sulphuric acid; and therefore acts as an energetic reducing agent. It is a dibasic acid, and forms acid, normal, and double sulphites.

EXPERIMENTS.—Sulphurous oxide is easily liquefied by passing it, previously thoroughly dried, through a U-tube immersed in ice and salt, as shown in Fig. 33. It may be preserved in sealed tubes, or if the quantity be large, in well stoppered mineral-water bottles. To show the cold produced by its evaporation, pour some of the liquid upon the surface of mercury contained in a capsule, and blow a current of air over it by means of a bellows. The mercury will be frozen. Or, pour some of the liquid acid into a thick crucible of platinum which is red-hot; the liquid will assume the spheroidal state at a temperature below its boiling point. If now, a little water be poured in, the sulphurous oxide will be instantly vaporized by the heat taken from the water, which therefore at once becomes ice. By some dexterity, the lump of ice may be thrown out of the red-hot crucible.

Fig. 33. Condensation of Sulphurous oxide.

The bleaching power of sulphurous oxide upon flowers may be illustrated by burning some sulphur under a glass shade, Fig. 34,

Fig. 34. Bleaching by SO_2

within which, on a tripod, are some brilliantly colored flowers. The flowers will be readily bleached, but at the same time will be very much wilted. That the color is not destroyed in these cases, may be proved very well by adding some sulphurous acid to two glasses, each of which contains some fresh infusion of the purple cabbage—an excellent vegetable color for testing acidity and alkalinity. The bleaching action is but slight until potassium hydrate solution is cautiously added, when the color entirely disappears, the two liquids becoming colorless. But if a little strong sulphuric acid be added to one, and a little potassium hydrate solution to the other, the color re-appears in both; in the first case, brilliant red, in the other brilliant green. Ether, benzol, and some other substances, will also restore the color of bodies thus bleached.

The deoxidizing power of sulphurous acid may be shown by adding its solution to one of potassium permanganate. The deep purple color of the latter solution at once disappears.

195. Tests and Uses.—Sulphurous oxide when free, is at once detected by its pungent odor, and by its blackening action upon paper moistened with a solution of mercurous nitrate. In combination as a sulphite, it evolves hydrogen sulphide when added to a solution evolving hydrogen.

In the arts it is used chiefly for bleaching straws and woolens; but for the reasons just given, the bleaching is not permanent as it is with chlorine, but requires to be frequently repeated. On account of its reducing power sulphurous acid is extensively used as a preserving fluid, both for cider, and, in the form of sodium sulphite, for canning fruits and vegetables.

SULPHURIC OXIDE.—*Formula* SO_3. *Molecular weight* 80. *Molecular volume* 2. *Density* 40. *One liter of the vapor weighs* 3·58 *grams* (40 *criths*).

196. Preparation.—Sulphuric oxide may be prepared by oxidizing sulphurous oxide. When this gas is mixed with oxygen, both being perfectly dry, and the mixed gases are passed over heated platinized asbestus, they unite, and form sulphuric oxide. It is also obtained by heating di-sulphuric acid :—

$$\underset{\textit{Di-sulphuric acid.}}{H_2S_2O_7} = \underset{\textit{Sulphuric acid.}}{H_2(SO_4)} + \underset{\textit{Sulphuric oxide.}}{SO_3}.$$

The vapor evolved is collected in a cold and dry receiver.

197. Properties.—Sulphuric oxide is a white, wax-like solid, crystallizing in silky fibers resembling asbestus. Its specific gravity is 1·9. It melts at 16°, and boils at 46°. On maintaining the temperature of the fluid oxide below 25°, it gradually changes into a solid, polymeric apparently with the one just mentioned, and called β sulphuric oxide ; this, at 50°, becomes fluid again, being transformed into the α form. The liquid oxide has the highest co-efficient of expansion of any known liquid, being 0·0027 for each degree. It unites with water with the evolution of great heat, producing sulphuric acid.

HYDROGEN SULPHATE OR SULPHURIC ACID.—*Formula* H_2SO_4. *Molecular weight* 98. *Specific gravity of liquid* 1·85. *Boils at* 325°.

198. History and Occurrence.—Sulphuric acid was prepared by Basil Valentine in the 15th century under the name "oleum sulphuris per campanum." Dr.

Roebuck proposed the present method of manufacture in 1770. The acid thus made is therefore called "English" acid.

Sulphuric acid occurs free in the waters of certain rivers and mineral springs. Boussingault estimates that the Rio Vinagre in South America carries daily to the sea more than 38,000 kilograms; and the water of the Oak Orchard mineral spring, New York, contains in each liter over 2½ grams. It has also been observed as a secretion of certain mollusks; the saliva of Dolium galea Lk. containing nearly 3½ per cent of it. In the sulphates of iron, calcium, barium, and strontium, forming the minerals melanterite, gypsum, barite, and celestite, sulphuric acid is also represented.

199. Preparation.—Sulphuric acid is prepared by adding water to sulphuric oxide, either at the instant of its formation or subsequently:—

$$H_2O + SO_3 = H_2SO_4.$$

When sulphur is boiled in nitric acid, it is oxidized to sulphuric oxide, which unites at once with the water present, forming sulphuric acid. In the preparation on the large scale, there are two general stages; 1st, the oxidation of the sulphurous to sulphuric oxide by the oxygen of the air; and 2d, the solution of the sulphuric oxide in water. The agent employed for carrying oxygen from the air to the sulphurous oxide is nitrogen tetroxide, N_2O_4. The entire process may be represented by the four following steps:—

1st. The burning of the sulphur to get sulphurous oxide:—

$$S + O_2 = SO_2.$$

2d. The reaction of the sulphurous and nitrogen oxides:—

$$(SO_2)_2 + N_2O_4 = (SO_3)_2 + N_2O_2.$$

3d. The union of the sulphuric oxide with water :—

$$SO_3 + H_2O = H_2SO_4.$$

4th. The re-oxidation of the N_2O_2 from the air :—

$$N_2O_2 + O_2 = N_2O_4.$$

In practice the operation is conducted in large leaden chambers, shown in Fig. 35 on the following page. The sulphur is burned in the furnace seen on the left, and the sulphurous oxide produced passes up through the large pipes, through a smaller and then a larger chamber, into a third, upon the floor of which, porous, earthen, terrace-shaped vessels are placed, over which a stream of nitric acid flows from the reservoirs just above. Here the sulphurous becomes sulphuric oxide and unites with the N_2O_2 produced to form a crystalline compound. The third stage, the union of the sulphuric oxide with water is effected by blowing steam into the chambers from a boiler heated by the burning sulphur, as shown in the figure. The sulphuric acid resulting from the decomposition of the crystalline compound and the union of its sulphuric oxide with water, collects on the floor of the chambers, while the N_2O_2 unites with the oxygen of the air present, to form N_2O_4, and thus renews the oxidation. A current of air passes slowly through these chambers, and to prevent loss, especially of the nitrogen oxides, the escaping products pass up through a column of moistened pumice, seen on the right, by which these are absorbed. The solution thus obtained is carried toward the sulphur furnace and allowed to trickle slowly over a series of inclined shelves in the first chamber, where it meets the entering sulphurous oxide and is utilized. A series of smaller chambers is generally pre-

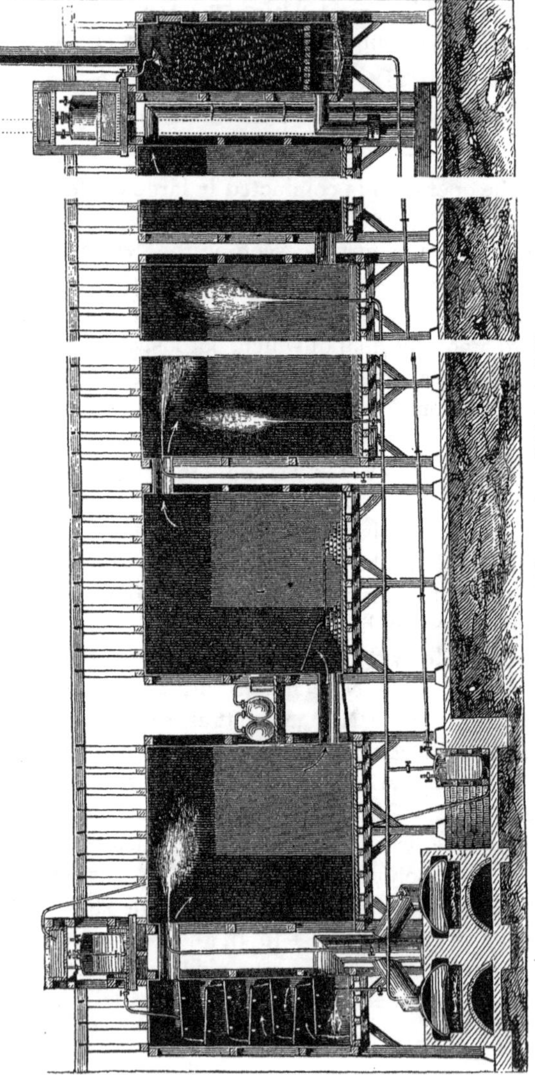

Fig. 35. Commercial Preparation of Sulphuric Acid.

ferred to a single large one. When the acid which accumulates in the water on the floor of the chambers has a specific gravity of 1·5, it is drawn off, concentrated in leaden vats by heat until the specific gravity rises to 1·7, and then in a platinum still or in retorts of glass, until all the water is driven off and the specific gravity rises to 1·85. It is then placed in carboys for use. In practice the leaden chambers often have a capacity of 100,000 cubic feet, and produce continuously thousands of tons per week. Sulphur itself is generally employed in this manufacture, though in some cases, the sulphurous oxide is obtained by roasting pyrite.

EXPERIMENTS.—To show the reducing action of sulphurous oxide upon nitrogen oxides, place in a jar filled with sulphurous oxide, Fig. 36, a stick dipped in strong nitric acid. Red fumes of the reduced nitrogen compounds will at once fill the jar, and soon unite with the sulphuric oxide to form a crystalline compound which lines the walls of the vessel. On adding water, the crystals dissolve with effervescence, the red fumes again appear, and sulphuric acid may be found in the liquid at the bottom of the jar.

Fig. 36. Oxidation of SO_2 by HNO_3.

Upon the lecture-table, the sulphuric acid process may be illustrated by the apparatus shown in Fig. 37. The lead-chamber is represented by the large glass globe, at first full of air. The two-necked bottle on the right contains the materials for generating N_2O_2; this gas enters the globe, meets with the air, and becomes N_2O_4. By means of a mixture of sulphur and manganese di-oxide contained in the flask shown on the left, sulphurous oxide is evolved and is led into the globe by the connecting tube. There meeting with the N_2O_4, the second reaction given above takes place, N_2O_4 and SO_2 becoming $(N_2O_2)SO_3$, which lines the walls of the globe, now colorless, with white radiating crystals. If finally, a jet of steam be blown in from the third flask, the crystals disappear, the globe becomes filled with red vapors, and sulphuric acid collects at the bottom. By renewing the air from time to time, through the rubber tube shown on the right, the process may be made continuous.

Fig. 37. Production of Sulphuric acid.

200. Properties.—Sulphuric acid is a dense, colorless, oily, and very corrosive acid liquid, having a specific gravity of 1·854. It boils about 338° and solidifies when cooled to 10·8°. It may be distilled, but suffers a partial decomposition, so that the product contains but 98·7 per cent of acid; this separation into sulphuric oxide and water,—or dissociation, as it is called—takes place completely at higher temperatures, so that its vapor-density is only one half of that required by theory. Sulphuric acid has a very strong attraction for water, combining with it with the evolution of great heat. It attracts moisture from the air, and is often used to dry a gas by causing it to bubble through the acid. It also removes water from organic matters placed in it, completely charring them.

Experiments.—To show the heat evolved by the union of sulphuric acid and water, pour one part of water upon four parts strong sulphuric acid in a beaker, and stir the mixture with a test-tube containing some ether or alcohol, colored with alkanet, or other coloring matter. The alcohol or ether will boil violently; by holding the tube in a stand, the vapor may be ignited, producing a voluminous flame.

To show its action on organic matters, add to 50 c. c. of sulphuric acid, an equal volume of strong sugar-syrup. Upon stirring the two together, the mass will become hot and rise into a black porous coal.

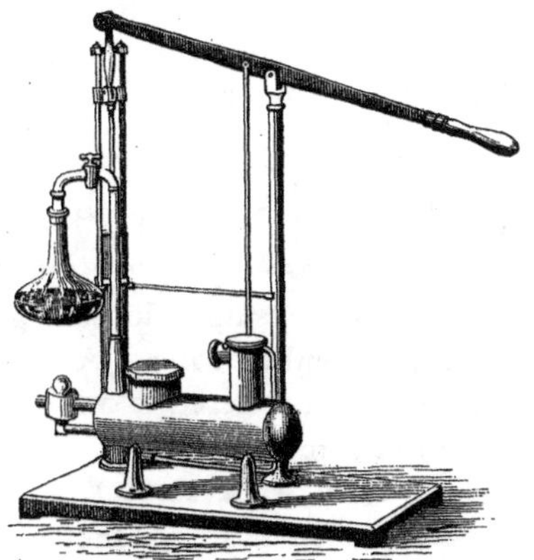

Fig. 38. E. Carré's Ice Apparatus.

The attraction of sulphuric acid for water is now made use of largely in Paris for the production of ice. Fig. 38 shows the apparatus contrived by Carré for this purpose. The water to be frozen is placed in the flask on the left, which is connected by a tube with a horizontal reservoir containing sulphuric acid; this reservoir may be exhausted by the air pump, the sulphuric acid being continually agitated. The water is cooled by its own evaporation under the diminished pressure; and as the vapor produced is at once removed by the sulphuric acid, it soon congeals. A pint of water may be frozen in 15 seconds with this apparatus.

The sulphuric acid which has now been considered, is the di-meta form, according to the previous classification of acids, page 36. By limiting the temperature during evaporation to 205°, by cooling a mixture of acid and water of specific gravity 1·78, or by mixing together 100 parts of the acid and 18·4 parts of water, an acid is obtained having the composition $H_4S^{VI}O_5$, which is mono-meta-sulphuric acid. It has a specific gravity of 1.78, and at 7° crystallizes in the rhombic system. Again, if a dilute acid be carefully evaporated at 100°, a third definite compound of water and sulphuric oxide results. It has the formula $H_6S^{VI}O_6$, and is ortho-sulphuric acid.

The common form of sulphuric acid is di-basic; sulphates may therefore be acid, normal, or double. Letting M stand for a monad metal, they may be represented thus :—

$$\left.\begin{matrix}H\\H\end{matrix}\right\}SO_4 \qquad \left.\begin{matrix}M\\H\end{matrix}\right\}SO_4 \qquad \left.\begin{matrix}M\\M\end{matrix}\right\}SO_4 \qquad \left.\begin{matrix}M\\M\end{matrix}\right\}SO_4$$

Acid. *Acid salt.* *Normal salt.* *Double salt.*

The other forms of sulphuric acid given above have also their corresponding salts. A zinc mono-meta-sulphate Zn''_2SO_5, and a mercuric ortho-sulphate Hg''_3SO_6 are well known compounds.

201. Di-sulphuric Acid, $H_2S_2O_7$. — Another kind of sulphuric acid is found in commerce, prepared by the distillation of dried ferrous sulphate in earthen retorts. It is a heavy oily liquid of specific gravity 1·9; it is usually more or less dark colored, hisses like a hot iron when dropped into water, and fumes strongly in the air. It is therefore called fuming sulphuric acid, or, as it is manufactured in Nordhausen in Saxony, sometimes Nordhausen sulphuric acid. The name di-sulphuric acid is given to it, because it may be regarded as derived from two molecules of sulphuric acid, by the removal of one molecule of water, thus :—

$$(H_2SO_4)_2 - H_2O = H_2S_2O_7.$$

When heated, it decomposes into sulphuric acid and sulphuric oxide, according to the equation :—

$$H_2S_2O_7 = H_2SO_4 + SO_3.$$

It is used for dissolving the indigo with which the celebrated Saxony blues are made.

Three sulphates have long been known under the name of vitriols, because like glass; zinc sulphate, or white vitriol; ferrous sulphate or green vitriol; and copper sulphate or blue vitriol. Because sulphuric acid was first prepared by distilling the second of these, it has received the name oil of vitriol.

202. Tests.—The test for free sulphuric acid is the charring it causes. A natural water containing this acid, if used to moisten paper, will char it completely on drying at 100°. In combination in a soluble form, sulphuric acid and sulphates give a dense white precipitate with solution of barium chloride, insoluble in acids. If the sulphate be insoluble in water, it may be recognized by fusing it with sodium carbonate, thus converting it into sodium sulphate. This is soluble in water and may be tested as above. Or, what is sometimes preferable, the suspected sulphate may be heated for some time with pulverized charcoal; it will thus be reduced to a sulphide, which, on treatment with a drop of acid, will evolve the well-known odor of hydrogen sulphide.

203. Uses.—Sulphuric acid is the most important substance consumed in chemical manufactures. It is used in the production of nitric, hydrochloric, phosphoric, citric, and tartaric acids; and in the manufacture of soda, of phosphorus, of alum, and of the alkaloids. It is largely used in dyeing, in calico printing, and in bleaching; in the preparation of fertilizers, and in the refining of petroleum. Indeed, the extent of the consumption of sulphuric acid by any nation, it has been well said, is a true index of its commercial prosperity.

THIONIC ACIDS.

Beside the acids now given, in which there is but one atom of sulphur in each molecule, there are others containing more than one such atom, the two or more sulphur atoms having different equivalences. This group of acids, called the thionic series, from the Greek *ϑεῖον*, sulphur, contains the following substances :—

Sulpho-sulphuric acid $H_2S_2O_3$

$$\begin{array}{ccccc} & & S & & \\ & & \| & & \\ HO & - & S & - & OH \\ & & \| & & \\ & & O & & \end{array}$$

Dithionic acid $H_2S_2O_6$

$$\begin{array}{ccccccc} & & O & & O & & \\ & & \| & & \| & & \\ HO & - & S & - & S & - & OH \\ & & \| & & \| & & \\ & & O & & O & & \end{array}$$

Trithionic acid $H_2S_3O_6$

$$\begin{array}{ccccccccc} & & O & & & & O & & \\ & & \| & & & & \| & & \\ HO & - & S & - & S & - & S & - & OH \\ & & \| & & & & \| & & \\ & & O & & & & O & & \end{array}$$

Tetrathionic acid $H_2S_4O_6$

$$\begin{array}{ccccccccccc} & & O & & & & & & O & & \\ & & \| & & & & & & \| & & \\ HO & - & S & - & S & - & S & - & S & - & OH \\ & & \| & & & & & & \| & & \\ & & O & & & & & & O & & \end{array}$$

Pentathionic acid $H_2S_5O_6$

$$\begin{array}{ccccccccccccc} & & O & & & & & & & & O & & \\ & & \| & & & & & & & & \| & & \\ HO & - & S & - & S & - & S & - & S & - & S & - & OH \\ & & \| & & & & & & & & \| & & \\ & & O & & & & & & & & O & & \end{array}$$

204. Sulpho-sulphuric Acid. — Hypo-sulphurous acid of some authors. Sulpho-sulphates are prepared either by adding sulphur to a sulphite or by partial oxidation of a sulphide. By the first method :—

$$Na_2SO_3 + S = Na_2S_2O_3$$

Large quantities of the sodium salt are manufactured for use in photography, under the name sodium hyposulphite. The acid corresponding to it has not been prepared in the free state.

SULPHUR AND CHLORINE.

205. The Sulphides of Chlorine are three in number, having the formulas Cl_2S_2, Cl_2S, and Cl_4S. They are formed by the direct union of their constituents, the first being formed when the sulphur is present in excess, the last when the chlorine is most abundant. The last, Cl_4S, is not known in the free state, but exists in combination with certain metallic chlorides as $SnCl_4(Cl_4S)_2$ with stannic chloride, $(SbCl_5)_2(Cl_4S)_3$ with antimonic chloride, etc.

§ 3. SELENIUM AND TELLURIUM.

SELENIUM.—*Symbol* Se. *Atomic weight* 79. *Molecular weight* 158. *Equivalence* II, IV, VI. *Density of vapor* 79. *Molecular volume* 2. *Specific gravity of solid* 4·5. 1 *liter of selenium-vapor at* 1420° *weighs* 7·08 *grams* (79 *criths*).

206. History and Preparation.—Selenium was discovered by Berzelius in 1817 in the lead-chamber deposits of the sulphuric acid manufactory at Gripsholm. He named it from σελήνη, the moon. It is a rare substance, occurring occasionally free, but generally combined with copper, lead, silver and mercury, as selenides. It exists also as an impurity in certain sulphurs. It is obtained from the sulphuric acid residues, or from the minerals containing it, by fusing with sodium nitrate and carbonate, extracting the sodium selenate with water, and reducing this with solution of sulphurous acid.

207. Properties.—Selenium like sulphur, is capable of existing in at least two allotropic states: α Selenium, which corresponds to α sulphur, is a dark grayish-black solid, of specific gravity 4·8, and is insoluble in carbon di-sulphide. β Selenium, is dark reddish-brown in color, has a specific gravity of 4·5, and is soluble in carbon di-sulphide, from which it crystallizes in monoclinic prisms. This is the more stable form of selenium, the form it has when native. A third or amorphous variety with a specific gravity 4·26 is also supposed to exist. On raising the temperature to 100°, β passes into α selenium, with a distinct evolution of heat. α selenium melts a little above 100°, and the β variety at 217°. The liquid boils at about 700°. In its chemical properties, selenium very closely resembles sulphur, forming similar compounds with other elements. It burns with a blue flame and gives often an intolerable odor like that of decaying horse-radish. It unites directly with the metals, forming selenides.

TELLURIUM.—*Symbol* Te. *Atomic weight* 128. *Equivalence* II, IV, *and* VI. *Density of vapor* 128. *Molecular volume* 2. *Specific gravity of solid* 6·2. 1 *liter of tellurium-vapor at* 1390° *weighs* 11·47 *grams* (128 *criths*).

208. History and Preparation.—Tellurium was discovered by Klaproth in 1798, in a Transylvanian gold ore, and named from tellus, the earth. It occurs even more rarely than selenium, being found native only in Hungary. It exists also in combination with bismuth, lead, gold, and silver. The mode of its preparation is analogous to that of selenium. It is obtained in solution either as potassium telluride, or as tellurous acid, and then precipitated; in the former case by a current of air, in the latter by sulphurous acid.

209. Properties.—Tellurium is a tin-white, brittle solid, having a strong metallic luster. It crystallizes in rhombohedrons, and conducts heat and electricity readily. It melts at 500°, volatilizes at a white heat, and may be distilled. Its vapor is greenish-yellow like chlorine. When heated in the air it takes fire and burns with a blue flame tinged with green, evolving white fumes of tellurous oxide. Indeed in all its physical properties it is a metal; but it is so closely allied chemically to sulphur and selenium, that it is considered with these elements. Its compounds are called tellurides.

PROPERTIES OF THE GROUP.

210. The same gradation of properties is seen here which was noticed in the chlorine group. As the atomic weight increases, the chemical activity diminishes. The sum of the atomic weights of sulphur and tellurium (32 + 128), is almost exactly double that of selenium, 79. They all form similar compounds with hydrogen, H_2O, H_2S, H_2Se, and H_2Te, in which they are bivalent; and the last three form oxides, in which their equivalence is four and six; as SO_2 and SO_3; SeO_2 and SeO_3; TeO_2 and TeO_3, to each of which there is a corresponding acid.

EXERCISES.

§ 1.

1. By what physical and chemical methods may oxygen be obtained?

2. What volume does a gram of oxygen occupy?

3. One gram of mercuric oxide yields what weight of oxygen?

4. How much mercuric oxide is required to yield 356 c. c. of oxygen measured at 15° and under 736 m. m. pressure?

5. What weight of oxygen measured at 100°, is necessary to fill a gas-jar which holds 4·6 liters of water?

6. At what temperature do 750 c. c. of oxygen occupy a liter?

7. Calculate the percentage of oxygen in CuO, MnO_2, $KClO_3$, KNO_3.

8. How much potassium chlorate is necessary to yield a cubic meter of oxygen? A kilogram?

9. If the chlorate be one dollar a kilogram, what will the oxygen cost per cubic meter?

10. A liter of oxygen is required of the density of 100 at 0°; how much $KClO_3$ is needed, and what is the pressure on the gas?

11. How much O will one liter of chlorine evolve from water?

12. From what is the name oxygen derived? Illustrate.

13. By what processes is oxygen obtained commercially?

14. When was ozone first recognized? By whom?

15. How may ozone be produced? What is Schönbein's test?

16. In what do oxygen, ozone, and antozone differ?

17. How is the composition of water proved by synthesis? By analysis?

18. What is the weight of a cubic meter of steam? What volume of hydrogen does it contain? Of oxygen?

19. One gram of water contains what volume of mixed gases?

20. If 226 c. c. of oxygen and 500 c. c. of hydrogen, both at 110°, be mixed and exploded, what will be the composition of the remaining gas, and what its volume at 0°?

21. What weight of potassium chlorate is needed to evolve the amount of oxygen contained in one c. c. of water?

22. What weight of water can be heated from 0° to 1° by the combustion of one cubic meter of mixed oxygen and hydrogen?

23. What volume has a block of ice which weighs a kilogram?

24. An iceberg floating in sea-water of specific gravity 1·027, exposes 30,000 cubic meters above the surface; what is its entire volume?

25. Define water of crystallization. Efflorescence. Deliquescence.

26. How is hydrogen peroxide prepared? What are the tests for it?

§ 2.

27. How does sulphur occur in nature? How is it purified?

28. Why is sulphur dimorphous? Prove its allotropism.

29. What do 632 c. c. of sulphur-vapor weigh at 500°? At 1000°?

30. One liter of hydrogen sulphide contains what weight of sulphur?

31. 500 c. c. H_2S requires what volume of oxygen for its combustion?

32. How many grams of FeS and of H_2SO_4 are needed to yield one cubic meter of H_2S? To saturate one liter of water at 0°?

33. Name the oxides and the acids of sulphur.

34. Sulphur burned in a liter of oxygen, gives what volume of SO_2?

35. Ten grams of S gives what volume of SO_2? What weight?

36. 53 grams copper yield how many c. c. of sulphurous oxide, measured at 100° and under 750 m. m. pressure?

37. To produce 100 grams calcium sulphite requires how much SO_2?

38. One kilogram of SO_3 requires the oxidation of what volume of SO_2?

39. What weight of $H_2S_2O_7$ will yield 100 c. c. of solid SO_3?

40. One gram of sulphur yields what weight of sulphuric acid?

41. Oil of vitriol of sp. gr. 1·773 contains 70 per cent of sulphuric acid; how many kilograms of such acid may be made from 150 kilograms of pyrite, containing 42 per cent of sulphur?

42. What are the chemical changes in the leaden chamber?

43. To neutralize a kilogram of lime requires what weight of H_2SO_4?

44. How is sulphuric acid detected? What salts does it form?

45. What is di-sulphuric acid and how is it made?

CHAPTER FOURTH.

NEGATIVE TRIADS.

§ 1. NITROGEN.

Symbol N. *Atomic weight* 14. *Equivalence* I, III, *and* V. *Density* 14. *Molecular weight* 28. *Molecular volume* 2. *One liter weighs* 1·25 *grams (*14 *criths).*

211. History.—Nitrogen was discovered by Rutherford in 1772. He showed that air, after it had been breathed by an animal and washed with lime-water, contained a gas which would support neither respiration nor combustion. Scheele and Lavoisier soon after showed independently, that this substance constituted four-fifths of the air. Lavoisier gave it the name azote, from α and ζωή. Chaptal proposed the name nitrogen, from νίτρον and γεννάω, because a necessary constituent of niter.

212. Occurrence.—Nitrogen exists free in the air, mixed with oxygen. It occurs also combined, in sodium, potassium and calcium nitrates and in ammonia. It forms an essential component of many vegetable and animal substances.

213. Preparation.—The easiest method of preparing nitrogen is to burn out of a given volume of air the oxygen it contains, thus leaving the nitrogen. It may also be procured by purely chemical processes ; as by heating ammonium nitrite :—

$$(NH_4)NO_2 = (H_2O)_2 + N_2,$$

or by passing chlorine gas through ammonia solution in excess :—

$$(H_3N)_8 + (Cl_2)_3 = (NH_4Cl)_6 + N_2.$$

EXPERIMENTS.—Nitrogen may be obtained from air by burning out the oxygen by phosphorus or copper. A fragment of phosphorus, carefully dried, is placed in a small capsule floated upon the surface of water by a piece of cork. The phosphorus is lighted and then covered with a large bell-glass, as shown in Fig. 39. Dense white fumes are formed by the combustion, which fill the jar; the oxygen is gradually consumed, and the water rises to takes its place. In a short time, these fumes disappear, and the nitrogen is left comparatively pure.

Fig. 39. Preparation of Nitrogen.

When copper is used, it is heated to redness in a glass tube, and a slow stream of air passed over it. The oxygen is retained by tne copper, and the nitrogen escapes from the tube.

For the chemical preparation of nitrogen, the ammonium nitrite —or what is equivalent to it, a mixture of ammonium chloride and potassium nitrite—is heated in an ordinary flask, and the gas is collected over water.

214. Properties.—I. PHYSICAL.—Nitrogen is a colorless, odorless, and tasteless gas, somewhat lighter than air, its specific gravity being 0·971. It has never been liquefied by cold or pressure. Water dissolves about

2·5 per cent of it. Its refractive power is to that of air as 1·034 to 1.

II. CHEMICAL.—Chemically, nitrogen is a remarkably inert substance, entering into direct combination with only a few elements, such as carbon, silicon, boron, and titanium, and, at an exceedingly elevated temperature, with oxygen. It extinguishes burning bodies introduced into it, and at ordinary temperatures, is not itself combustible. It is irrespirable, though it exerts no positively injurious action upon the tissues; animals die in it as they would in water, simply from suffocation. Though so indifferent when free, the compounds formed by nitrogen are among the most energetic known. The corrosive nitric acid, the pungent ammonia, the explosive nitro-glycerin, the active poisons known as prussic acid and the alkaloids, all contain nitrogen. Some chemists have long believed it to be compound.

THE ATMOSPHERE.

215. Physical Properties.—The atmosphere is the aerial envelope which surrounds the earth. Careful experiments by Regnault have shown that one liter of air weighs 1·2932 grams at 0°, and under 760 m.m. pressure; it is therefore 14·45 times heavier than hydrogen, and is the standard of specific gravity for gases. Torricelli showed in 1643, that the pressure of the air upon the earth's surface would sustain a column of mercury about 76 centimeters in hight; and as a column of mercury of this hight, whose area is one square centimeter, weighs 1033 grams, it follows that this number represents the atmospheric pressure upon every square centimeter of the earth's surface. From this it appears that the entire weight of the air on our globe is equal to that of a sphere of lead 100 kilometers in diameter. The hight of the atmosphere is unknown; it is generally given as 50 or 60 kilometers, but observations upon the zodiacal light and upon meteoric showers prove that it may be from 320 to 340 kilometers in hight. As we rise from the earth, the density of the air diminishes rapidly, according to Marriotte's law; so that half of it is within four and one-half kilometers of the surface. The barometer shows that the weight of the air fluctuates within narrow limits, this instrument varying sometimes as much as 60 millimeters in hight.

216. Chemical Properties.—Air is a mixture of oxygen and nitrogen gases. This may be ascertained both by

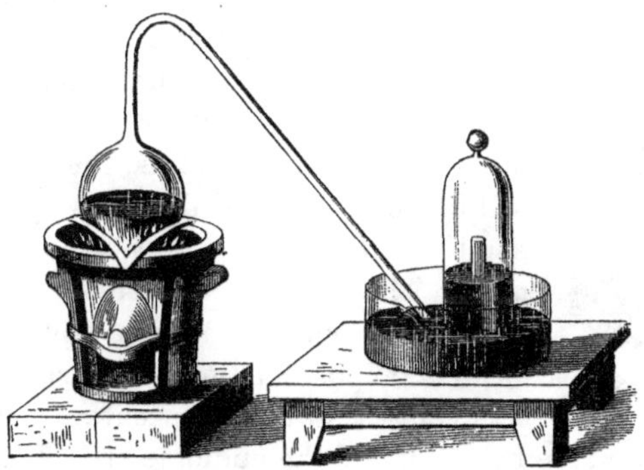

Fig. 40. Lavoisier's Experiment.

analysis and by synthesis. The former method is the one by which Lavoisier first established the composition of the air. His experiment, now a classic one in chemistry, was thus performed: a glass balloon with a long neck, bent as shown in Fig. 40, was partially filled with mercury and placed on a furnace. The neck passed down under the surface of the mercury in an adjoining vessel, and then up into a bell-glass—also full of air—whose mouth was sealed by the mercury. On raising the temperature of the mercury to near the boiling point, a red powder began to accumulate upon its surface, the volume of the air proportionally diminishing; until at the end of twelve days, the contraction of volume ceased, and the experiment was concluded. The gas contained in the apparatus was proved to be nitrogen; and by collecting the red powder and heating it, as in Fig. 17, the mercury was reproduced and a gas evolved which had all the properties of oxygen.

This experiment was qualitative; an approximate quantitative experiment may be made by taking a graduated tube full of air, placing in it a ball of phosphorus cast on the end of a wire, Fig. 41, and immersing its mouth in mercury. By the slow combustion of the phosphorus, the oxygen will be removed, and the mercury will rise to fill the space it previously occupied. The nitrogen will be left in the tube. Knowing the original volume of air, its composition may be easily calculated. A still more accurate analysis may be made by means of the eudiometer. Fig. 42 represents a convenient form for the lecture-room, known as Volta's eudiometer. It consists of a strong cylinder of glass, closed above and below by stop-cocks, the lower one carrying a funnel for convenience of filling, the upper one a cup for holding water, into which may be screwed the long graduated tube, shown in the figure.

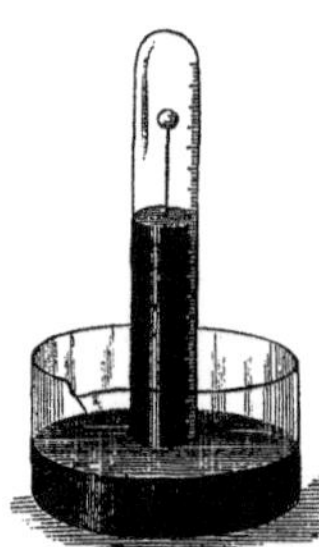

Fig. 41. Analysis of Air by Phosphorus.

To make an analysis of air, a given portion, say 200 c. c., is introduced into the eudiometer—previously filled with water and standing on the water-cistern — by means of the measuring glass shown on the right. Sufficient hydrogen to combine with all the oxygen, say 100 cubic centimeters, is then added, the lower cock is closed, and an electric spark passed through the mixture from the small ball attached to the upper cap. The hydrogen and oxygen unite, and, on opening the lower cock, the water will enter to take the place of the gas which has disappeared. The long graduated tube is now filled with water, inverted into the top cup of water, and screwed to its place. The top cock is now opened, and, by depressing the apparatus in the cistern, the remaining volume of gas will pass into this tube and may be measured: assuming that it measures 174 cubic centimeters, then the volume of gas which has disappeared must be 300 −174 or 126 cubic centimeters. But this 126 c. c. must be two-thirds hydrogen and one-third oxygen, this being the ratio in which these two gases combine by volume. One third of 126 is 42; hence 200 c. c. of air contain 42 c. c. of oxygen, and 100 volumes contain 21 volumes of oxygen.

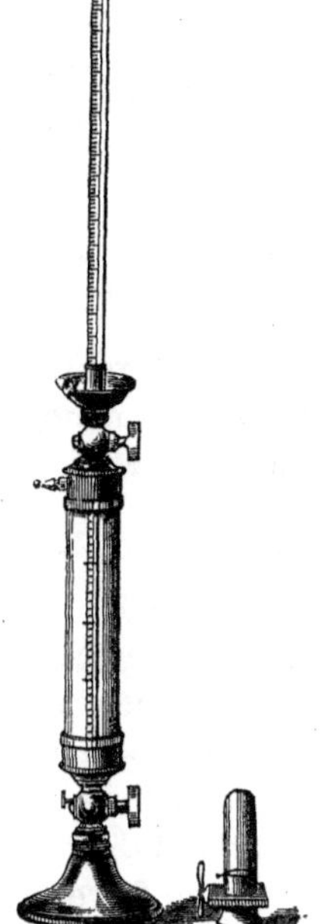
Fig. 42. Volta's Eudiometer.

The most accurate analyses of air on record were

made by Dumas and Boussingault, by drawing pure dry air over red-hot copper. The increase in the weight of the copper gave the weight of the oxygen, and the increased weight of the exhausted globe, that of the nitrogen drawn into it. In this way the composition of the air by weight was directly, and by volume indirectly, determined to be as follows:—

	By weight.	*By volume.*
Oxygen, . . .	23·0	20·8
Nitrogen, . . .	77·0	79·2
	100·0	100·0

The air of different localities, though nearly constant in composition, is not absolutely so; the oxygen may diminish from 21 volumes to 20·9, and in rare cases, even to 20·3.

That the air is merely a mechanical mixture of its constituents, and not a chemical compound, is proved by the following considerations:—1st, its components are not united in the ratio of their atomic weights; 2d, the properties of air are such as might properly be expected of a mixture; 3d, each gas dissolves in water independently of the other; and 4th, no change of volume or evolution of force appears when air is made artificially by placing together oxygen and nitrogen.

EXPERIMENTS.—A mixture of oxygen and nitrogen, in the proportion of one volume of the former gas to four of the latter, made in a jar over the water-cistern, acts, in reference to combustible bodies, precisely like common air. The contrast between air and its constituents may be shown by taking three jars, one of nitrogen, one of oxygen, and a third of the artificial air, made as above, and introducing into them successively, a lighted taper with a long wick. In the first jar it will be extinguished, in the second—provided a spark is left on the wick—it will be relighted, and in the third it will burn normally, as in the outside air.

Water, on being boiled, loses the air which it has dissolved. On collecting and analyzing this air it is found to be richer in oxygen than common air, having 32 per cent of this gas and 68 of nitrogen. As the coefficient of solubility of both these gases is known, it is easy to calculate what the composition of the dissolved gases should be, on the the supposition that the air is a mixture. Calling the air one-fifth oxygen and four-fifths nitrogen, and the coefficient of solubility of oxygen ·046 and of nitrogen ·025, we have :—

	Solubility calculated.		*Solubility observed.*
Oxygen	$·046 \times \frac{1}{5} =$ ·0092	or 31·5	32
Nitrogen	$·025 \times \frac{4}{5} =$ ·0200	or 68·5	68
	·0292	100·0	100

This correspondence establishes the fact of mixture, since every chemical compound has a specific solubility of its own. This larger percentage of oxygen in the air dissolved by water, it may here be observed, is essential to the life of fishes.

The density of oxygen being to that of nitrogen as 16 : 14, it might be expected that they would separate, the heavier oxygen accumulating near the earth. But we have seen that all molecules are in constant motion ; and hence that all gases readily permeate or diffuse into each other independently of their density. The perfection of this diffusion is shown in the fact that the variation in the composition of the air is as slight as analysis has showed it to be.

Fig. 43. Properties of N and O contrasted.

EXPERIMENT.—The relative densities of oxygen and nitrogen, as well as their opposite action upon flame, may be well shown by the apparatus given in Fig. 43. Two bell-glasses are filled, the one with oxygen, the other with nitrogen, closed by plates of glass and placed together, the oxygen lowest, as shown in the cut. On removing the stopper of the upper jar and the plates between the two, and introducing a lighted taper having a long wick, the flame is extinguished in the nitrogen but relighted again—if a spark be left on the wick—as it descends into the oxygen. This may be repeated several times before the gases become mixed by diffusion.

But beside these two chief components of the air, it contains other substances in small quantity, which are quite as essential. These are aqueous vapor, carbonic di-oxide, and ammonia. The aqueous vapor varies very widely in amount, depending on various conditions, especially on temperature. The quantity of moisture in the air is measured by the hygrometer; air is said to be saturated when it contains all the moisture it can hold at any given temperature; thus at 0°, one cubic meter is saturated by 5·4 grams of water, at 10° by 9·74 grams, and at 25° by 22·5 grams. But the air is seldom entirely saturated; 60 per cent is regarded as the healthy mean; but it may contain only one-fifteenth of the saturating quantity, as is the case on the Red Sea during a simoon. When the air is cooled, the excess of moisture falls as rain; thus one cubic meter at 25° cooled to 10° would deposit 22·5—9·74 or 12·76 grams of water. The carbonic di-oxide of the air, the next largest constituent, exists in minute quantity relatively—about one-twentieth of one per cent—though the absolute quantity is large, being about 3,000 billion kilograms. It is estimated by drawing a known volume of air through a tube containing potassium hydrate, which absorbs it and thus increases in weight. This minute amount of carbonic di-oxide is the sole source of the carbon of vegetation. It is produced by combustion, by the respiration of animals, by fermentation, and by decay. The ammonia present in air exists in even more minute quantity, being only from one to fifty parts in a million of air, according to the locality. This ammonia washed down by the rain, plays an important part in yielding nitrogen to vegetation. Other and variable constituents there are in the air such as organic matters of various kinds, dust, and gaseous products. Miller gives the average composition of the air of England, as follows:—

Oxygen	20·61
Nitrogen	77·95
Carbonic di-oxide . . .	·04
Aqueous vapor . . .	1·40
Nitric acid / Ammonia / Gaseous hydrocarbons . .	traces.

And in towns:—

Hydrogen sulphide / Sulphurous oxide . .	traces.
	100·00

NITROGEN AND HYDROGEN.

HYDROGEN NITRIDE OR AMMONIA.—*Formula* H_3N. *Molecular weight* 17. *Molecular volume* 2. *Density* 8·5. *One liter weighs* 0·762 *grams (*8·5 *criths).*

217. History.—Ammonia was known to the alchemists; it was mentioned by Raymond Lully in the 13th century, and by Basil Valentine in the 15th. It was first obtained as a gas by Priestley in 1774, and called alkaline air. Scheele in 1777 showed that it contained nitrogen, and Berthollet analyzed it in 1785. The name ammonia was given by Bergman in 1782, from that of its chloride, then called sal-ammoniac; which substance was largely produced by burning camel's dung in the Libyan desert, near a temple of Jupiter Ammon. It occurs sparingly in nature, traces of it being found in the air, in soils, and in most mineral waters. It exists also in certain minerals found in volcanic regions, and in the fluids of animals and plants.

218. Preparation.—Ammonia cannot be produced by the direct union of its constituents; though by suitable means they may be made to combine indirectly. When for example, nitric acid acts upon zinc, the hydrogen which is set free in the atomic form and hence with a stronger than molecular attraction—hence called nascent hydrogen—unites at once with the nitrogen, according to the following equation:—

$$\underset{\textit{Nitric acid.}}{(HNO_3)_9} + \underset{\textit{Zinc.}}{Zn_4} = \underset{\textit{Zinc nitrate.}}{(Zn(NO_3)_2)_4} + \underset{\textit{Water.}}{(H_2O)_3} + \underset{\textit{Ammonia.}}{H_3N}.$$

The compounds of ammonia found in commerce are obtained either by the destructive distillation of animal matters, or from the so-called ammoniacal liquors of the gas-works, obtained in the distillation of coal. Ammonia itself is prepared by acting upon two parts of ammonium chloride,—the sal-ammoniac of commerce—with one part of quick-lime. The reaction is as follows:—

$$(NH_3HCl)_2 + CaO = CaCl_2 + H_2O + (H_3N)_2.$$

The gas being lighter than air, may be collected by upward displacement.

219. Properties.—Ammonia is a colorless gas with a pungent odor and a strongly alkaline reaction upon test-papers. It is considerably lighter than air, its specific gravity being 0·59. Subjected to a pressure of six and a half atmospheres at 10°, or to a cold of—40°, it condenses to a colorless liquid of specific gravity 0·76 which freezes at—75°. It is soluble in water to an extraordinary degree ; one volume of water at 0° absorbing 1149 volumes of ammonia gas, forming the so-called aqua ammoniæ. At 15° one volume of water absorbs 783 volumes of the gas. This solution has the well-known properties of the gas, has a specific gravity of 0·85, and evolves the ammonia again upon heating. For the preparation of the ammonia solution, the same apparatus may be employed as was used for hydrochloric acid, Fig. 11.

Chemically, ammonia gas has a strong, but transient alkaline reaction upon vegetable colors, whence the name volatile alkali, sometimes applied to it. Though containing so much hydrogen it is not combustible in air at ordinary temperatures, though it burns in oxygen. A burning candle immersed in the gas is extinguished and animals die in it at once, owing to the extreme irritation it causes. Under the influence of heat or of the electric spark it is decomposed. As it contains trivalent nitrogen it can unite directly with other bodies, the nitrogen then becoming a pentad.

EXPERIMENTS.—The indirect formation of ammonia may be very beautifully shown by mixing in a gas-holder five volumes of hydro-

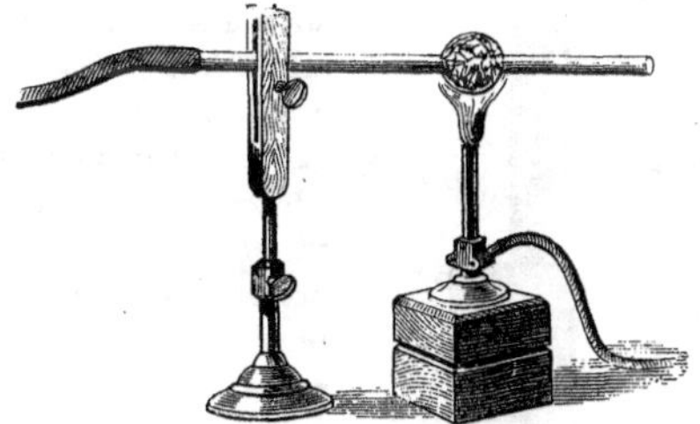

Fig. 44. Synthesis of Ammonia.

gen and two volumes nitrogen di-oxide, and passing a stream of the mixed gases through a bulb tube containing platinized asbestus, as shown in Fig. 44. So long as the bulb is cold, the escaping gases redden blue litmus paper ; but on warming the bulb, the surface action of the platinum begins, the asbestus often becoming red-hot, and the pungent alkaline fumes of ammonia appear, and turn the red paper back again to blue.

The absorption of ammonia gas by water may be illustrated by filling a large bottle with the gas by upward displacement, and closing the mouth with a rubber cork through which a glass tube passes, drawn to a fine point at the lower end. On breaking this point beneath the surface of water as shown in Fig. 45, the water will enter the bottle with great violence, sometimes crushing it. If the water be colored with red litmus solution, it will become blue as it enters the bottle, thus showing at the same time the alkalinity of the gas.

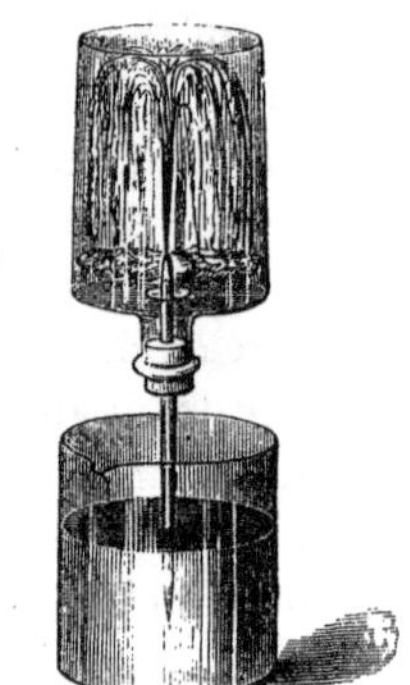

Fig. 45. Absorption of Ammonia.

The facility with which ammonia gas may be expelled from its solution by heat, and the ease with which it may be condensed to a liquid by pressure, have been made use of by F. Carré of Paris, for the production of artificial ice. His apparatus is represented in Fig. 46. It consists of a generator and receiver made of iron boiler-plate, the receiver being conical in shape, both connected by means of a strong iron tube. In the generator is placed a strong solution of ammonia saturated at 0°, and this is heated over a large gas flame, the receiver meanwhile being immersed in cold water. The ammonia gas is driven off and is condensed to the liquid state in the receiver, as soon as the pressure passes ten atmospheres. Into the cylindrical space in the receiver a closely-fitting vessel filled with water is now placed, and the apparatus is reversed, the generator being immersed in the water. The liquefied am-

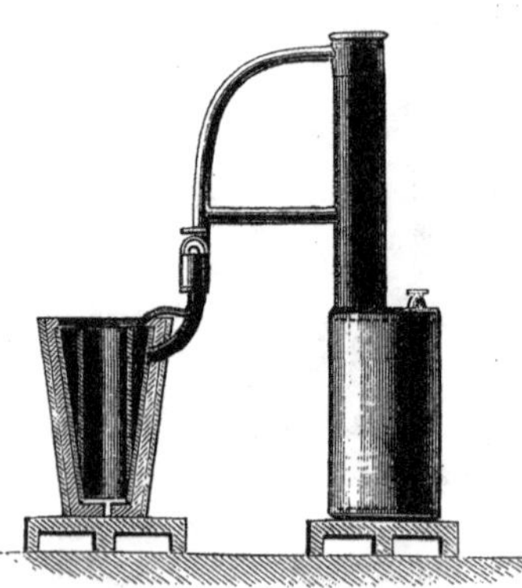

Fig. 46. F. Carré's Ammonia Ice-machine.

monia passes again into the gaseous state and is re-absorbed by the water in the generator. But in this evaporation great cold is produced and the vessel of water is soon frozen. A larger and continuous apparatus on the same principle, has also been patented by M. Carré.

The combustion of ammonia in oxygen may be conveniently shown by the apparatus represented in Fig. 47. The gas is obtained by heating a strong solution of ammonia in the retort, and is conducted through a narrow glass tube to a point just at the upper edge of a narrow glass cylinder, through which passes a current of oxygen, supplied by the flexible tube. The jet of ammonia gas as it issues, being surrounded by an atmosphere of oxygen, takes fire on the approach of a lighted taper, and burns with a peculiar yellowish flame.

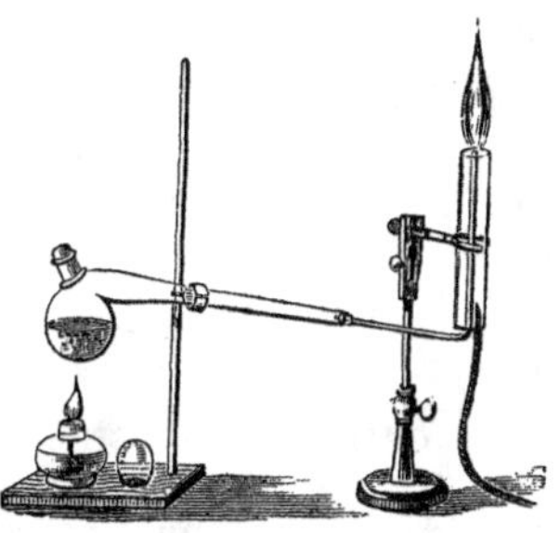

Fig. 47. Combustion of Ammonia in Oxygen.

220. Composition.—The composition of ammonia may be determined by introducing a given volume of the gas into a graduated tube over mercury, and passing electric sparks through it. It is decomposed and doubles in volume; the pungency and alkalinity of the gas disappear and it is no longer soluble in water. By eudiometry, the volumes of its constituents are obtained, thus: assuming that 100 cubic centimeters of ammonia are taken, they will expand to 200 c. c. on passing the spark; 100 c. c. of oxygen are now added and the spark again passed; the 300 c. c. become reduced to 75 c. c.; 225 c. c. having disappeared. Of this 225 c. c. two thirds or 150 c. c. must be hydrogen; and 75 c. c. oxygen. Subtracting the excess of oxygen taken, 100—75 or 25 c. c., from the residual 75 c. c. left in the eudiometer, the remainder, 50 c. c. is nitrogen. Hence the 200 expanded volumes consist of 150 volumes of hydrogen and 50 of nitrogen; and ammonia gas consists of three volumes of hydrogen and one volume of nitrogen condensed into two volumes.

221. Tests.—Free ammonia is easily detected by its odor, by its alkalinity, and by the fumes which it gives when a rod moistened with hydrochloric acid is brought near it. When combined, it may be set free by a little quicklime and then tested.

OXIDES AND ACIDS OF NITROGEN.

The oxides formed by nitrogen are five in number; those normally formed, in which it has an equivalence of one, three, and five; and those in which the nitrogen atoms are directly united, and which may be viewed as free radicals. Their names, together with their corresponding acids, are as follows:—

Oxides.		*Acids.*	
Hyponitrous oxide	N'_2O	Hyponitrous acid	$HN'O$
Nitrogen di-oxide (nitrosyl)	N'''_2O_2		
Nitrous oxide	N'''_2O_3	Nitrous acid	$HN'''O_2$
Nitrogen tetr-oxide (nitryl)	$N^V{}_2O_4$		
Nitric oxide	$N^V{}_2O_5$	Nitric acid	HN^VO_3

NITRIC OXIDE.—*Formula* N_2O_5. *Molecular weight* 108.

222. History and Preparation.—Nitric oxide, called also nitric anhydride, was first obtained by Deville in 1849. It is prepared by the action of chlorine, or better of phosphoryl chloride, upon silver nitrate, at 60°.

223. Properties.—Nitric oxide is a colorless, transparent solid, crystallizing in right rhombic prisms. It melts at 30° and boils at 47°. It is quite unstable, sometimes exploding spontaneously. It reacts energetically with water, producing nitric acid, thus:—

$$N_2O_5 + H_2O = (HNO_3)_2.$$

HYDROGEN NITRATE, OR NITRIC ACID. — *Formula* HNO_3. *Molecular weight* 63. *Molecular volume* 2. *Density* 31·5. *One liter of nitric acid vapor weighs* 2·82 *grams (*31·5 *criths).*

224. History.—Nitric acid was known to Geber, an alchemist of the 8th century; Raymond Lully in 1225 described a method for preparing it. Cavendish, in 1785, first determined its true composition synthetically.

225. Formation.—When strong electric sparks are passed through a confined portion of air, standing over a solution of potassium hydrate, the volume gradually lessens and potassium nitrate may be detected in the liquid. So when ozone acts upon the nitrogen of the air, upon ammonia, or upon the lower oxides of nitrogen, water being present, nitric acid is produced. Again, when animal matters containing nitrogen are allowed to decompose in presence of weak alkaline bases, nitrates of these bases are produced. In this way artificial niter-beds are made.

226. Preparation.—Nitric acid is always produced by the distillation of a nitrate—generally sodium or potassium nitrate—with sulphuric acid. The reaction may thus be represented:—

$$\underset{\text{Sodium nitrate.}}{Na(NO_3)} + \underset{\text{Sulphuric acid.}}{H_2(SO_4)} = \underset{\text{Nitric acid.}}{HNO_3} + \underset{\text{Hydro-sodium sulphate.}}{HNa(SO_4)}.$$

EXPERIMENTS.—The distillation of nitric acid may be conducted in the apparatus given in Fig. 48. The sodium nitrate is placed in the retort on the right and upon it is poured, through the tubulure, by means of a funnel, an equal weight of sulphuric acid. The neck

Fig. 48. Preparation of Nitric acid.

of the retort passes into that of the receiver for a considerable distance, and the receiver, supported over a beaker, is covered with paper to distribute equally the water which runs from the vessel above, and which is intended to keep it cool. On lighting the

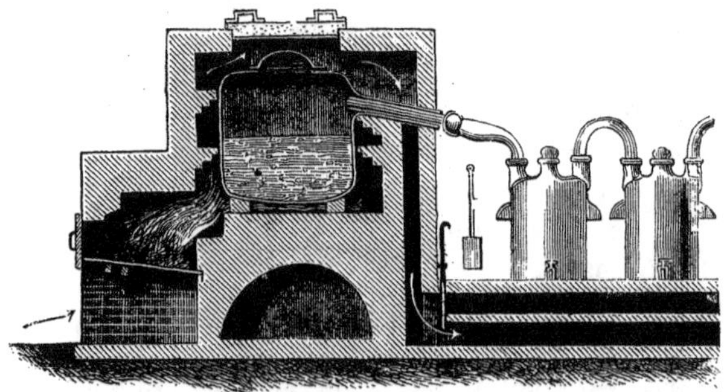

Fig. 49. Commercial Preparation of Nitric acid.

burner, the mass liquefies, red fumes appear, and a more or less colored liquid accumulates in the receiver. By changing this, collecting the acid which comes over during the middle of the operation separately, a colorless acid is obtained.

In the arts the operation is conducted in a cast-iron retort as shown in Fig. 49. A less concentrated acid is used for the decomposition, and two molecules of sodium nitrate are treated with one of sulphuric acid, normal sodium sulphate remaining in the retort thus :—

$$(Na\,NO_3)_2 + H_2(SO_4) = Na_2(SO_4) + (HNO_3)_2.$$

The nitric acid distills over and is condensed in the earthen-ware receivers.

227. Properties.—Nitric acid is a colorless, fuming, corrosive, strongly acid liquid, having a specific gravity of 1·52. Cooled to −55° it freezes, and heated to 86°, it boils, suffering a partial decomposition. It is also readily decomposed by light. Chemically, it is a powerfully oxidizing agent, acting on most of the metals with great vigor. Nitrogenous animal substances, such as parchment, silk, and wool, are colored strongly yellow by nitric acid. And many non-nitrogenous vegetable substances, such as glycerin, cotton, and sugar, are converted by it into violently explosive bodies.

The commercial acid—known as aqua-fortis—is of two sorts, called single and double. Double aqua-fortis has a specific gravity of 1·36, and single of 1·22, being one-half as strong. A mixture of nitric acid and water of density 1·42 has a definite boiling point, 120·5°. But it is not a definite hydrate, the boiling point being uniform only under a constant barometric pressure.

Nitric acid is a monobasic acid, and can form only normal salts, represented by $M(NO_3)$, M being any monad metal. But beside this, which is the di-meta-nitric acid, two others are possible, represented by H_3NO_4, mono-meta-nitric acid, and H_5NO_5, ortho-nitric acid. Lead mono-meta-nitrate $Pb''_3(NO_4)_2$, and bismuth mono-meta-nitrate, $Bi'''NO_4$—usually called basic nitrates—are well known salts ; and a hydro-bismuthous ortho-nitrate, $H_2Bi'''NO_5$, has also been produced.

228. Tests.—When free, nitric acid reddens litmus powerfully, bleaches indigo-solution readily, and evolves red fumes on introducing a fragment of copper. These reactions are obtained from nitrates after treatment with sulphuric acid. Moreover, nitrates deflagrate when thrown on burning charcoal.

Nitric acid is used in the arts for etching upon metals, for oxidizing various substances, for forming various substitution products such as nitro-benzol and picric acid, and for the preparation of nitro-glycerin, gun-cotton, etc.

229. Aqua Regia. — Neither nitric nor hydrochloric acid alone has the power of dissolving gold, the rex metallorum of the ancients. But a mixture of one volume of nitric and three of hydrochloric acid contains free chlorine and possesses this property, whence its name aqua regia. On submitting this mixture to a gentle heat, two exceedingly volatile liquids are obtained; one of these is chloro-nitryl, $(N_2O_2)Cl_4$, the other nitrosyl chloride $(NO)Cl$. Neither of these attack gold.

NITROGEN TETR-OXIDE, *or free Nitryl. Formula* N_2O_4 *or* $(NO_2)_2$. *Molecular weight* 92. *Molecular volume* 4 *(Dissociation).*

230. Preparation.—Nitrogen tetr-oxide is generally prepared by heating perfectly dry lead nitrate:—

$$(Pb''(NO_3)_2)_2 = (PbO)_2 + (N_2O_4)_2 + O_2.$$

On passing the vapors through a freezing-mixture, they are condensed to a liquid, or if perfectly dry, to a white crystalline solid which melts at $-9°$. As the temperature rises, the color of the liquid changes from yellow to deep orange, until it reaches 22° when it boils, evolving an orange vapor which at 40° is almost black. By the action of water it yields nitric and nitrous acids :—

$$\left.\begin{matrix}NO_2\\NO_2\end{matrix}\right\} + \left.\begin{matrix}H\\H\end{matrix}\right\}O = \left.\begin{matrix}NO_2\\H\end{matrix}\right\}O + \left.\begin{matrix}NO\\H\end{matrix}\right\}O.$$

It is an energetic oxidizing agent, being used to oxidize sulphurous oxide in the sulphuric acid process. It combines directly with chlorine to form nitryl chloride $(NO_2)Cl$.

NITROUS OXIDE AND ACID.—*Formulas* N_2O_3 *and* HNO_2. *Molecular weight of the oxide* 76 ; *of the acid* 47.

231. Preparation.—Nitrous oxide may be prepared by the reduction of nitric acid by starch or by arsenous oxide :—

$$(HNO_3)_2 + As_2O_3 = (HAsO_3)_2 + N_2O_3.$$

By passing the evolved vapors through a freezing mixture, the nitrous oxide condenses to a very unstable blue liquid, which reacts with water, producing nitrous acid ; this is also a blue liquid which may be preserved at low temperatures unaltered, but is decomposed readily into nitric acid, water and nitrogen di-oxide, thus :—

$$(HNO_2)_3 = HNO_3 + H_2O + N_2O_2.$$

Nitrous acid forms salts called nitrites ; the mono-meta form is monobasic, the ortho form, tribasic. Potassic mono-meta-nitrite is KNO_2 ; this is the more common form of nitrite. Hydro-plumbic ortho-nitrite $HPb''NO_3$ and normal plumbic ortho-nitrite $Pb''_3(NO_3)_2$ are examples of actually known ortho-salts.

EXPERIMENT.—The formation of nitrous compounds by the oxidation of ammonia may be illustrated by the apparatus shown in

Fig. 50. A flask is one-third filled with strong ammonia solution, and placed in a cup of sand on the gas furnace. A spiral of platinum wire one-third of a millimeter thick — formed by winding it about a pencil — is attached to a cork, heated to redness and plunged at once into the flask. At the same time, oxygen gas is admitted through a glass tube which just dips into the liquid. The spiral glows brilliantly in the gaseous atmosphere, producing at first white fumes of ammonium nitrite and then red vapors of nitrous oxide. When the ammonia gas is freely evolved, it forms often an explosive mixture with the oxygen; this ignited by the coil, gives a slight puff. The coil cooled by this explosion, soon again becomes heated and the operation is repeated. Sometimes the explosions are minute and take place rapidly within the flask producing a tone like that of the hydrogen tube.

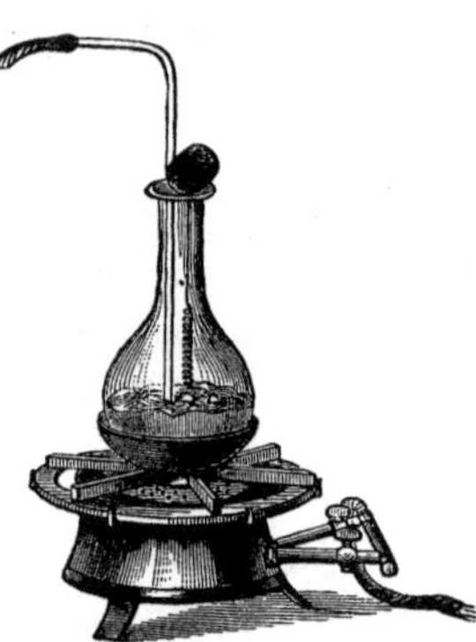
Fig. 50. Combustion of NH_3 to N_2O_3.

NITROGEN DI-OXIDE. — *Formula* N_2O_2. *Molecular weight* 60. *Molecular volume* 4 *(Dissociation). Density* 15. *One liter weighs* 1·34 *grams (* 15 *criths).*

232. History. — Nitrogen di-oxide, though noticed by Hales, was first investigated by Priestley in 1772.

233. Preparation.—It may be prepared by the reduction of nitric acid by metals, such as copper, silver, or mercury, or by ferrous sulphate. With copper, the reaction is as follows :—

$$Cu_3 + (HNO_3)_8 = (Cu(NO_3)_2)_3 + (H_2O)_4 + N_2O_2.$$

234. Properties.—Nitrogen di-oxide is a colorless and incoercible gas, having a specific gravity of 1·039.

There is an anomaly in its vapor-density, since its molecule if saturated, occupies four volumes instead of two. This is explained by the supposition that even at ordinary temperatures, the molecule is separated into two others, each of which occupies the normal volume. This is known to be the case with some other bodies, alike irregular in their vapor-density. One volume of this gas dissolves in about twenty of water at 15°.

Nitrogen di-oxide extinguishes the flame of a candle introduced into it; but phosphorus well ignited, burns in it with great brilliancy, being able to take away its oxygen. It has a strong attraction for oxygen, combining with half its volume of this gas to form red fumes of nitrogen tetroxide, N_2O_4. It also unites directly with chlorine, producing nitrosyl chloride (NO)Cl.

EXPERIMENTS.—To prepare nitrogen di-oxide, nitric acid of specific gravity 1·2—prepared by diluting the ordinary acid with twice its volume of water—is poured upon copper clippings contained

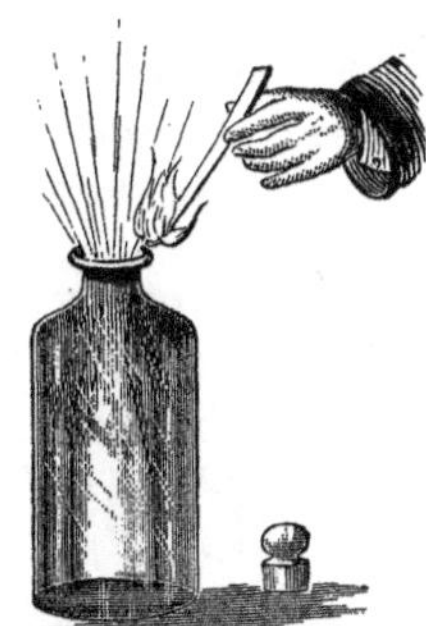

Fig. 51. Combustion of CS_2 and N_2O_2.

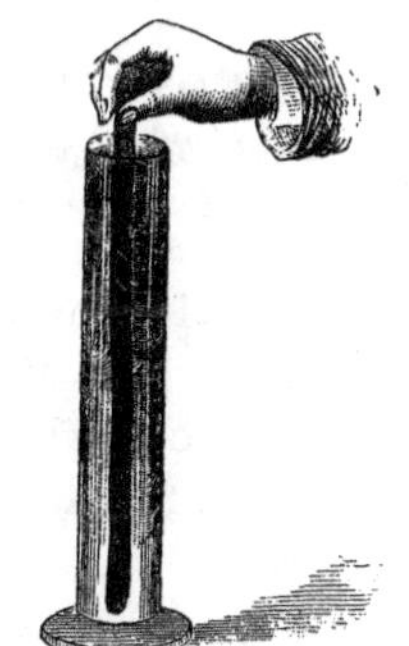

Fig. 52. Action of N_2O_2 and N_2O_3 on Litmus Paper.

in a two necked bottle like that used for obtaining hydrogen, Fig. 2. The gas must be collected over water.

A lighted candle or burning sulphur is extinguished when

plunged into the gas. Phosphorus just ignited is also extinguished; but if allowed to get fully on fire it burns brilliantly. If a few drops of carbonic disulphide be poured into a jar of the gas and agitated, the mixture will take fire on the approach of a flame as shown in Fig. 51, burning with a vivid, intensely actinic light.

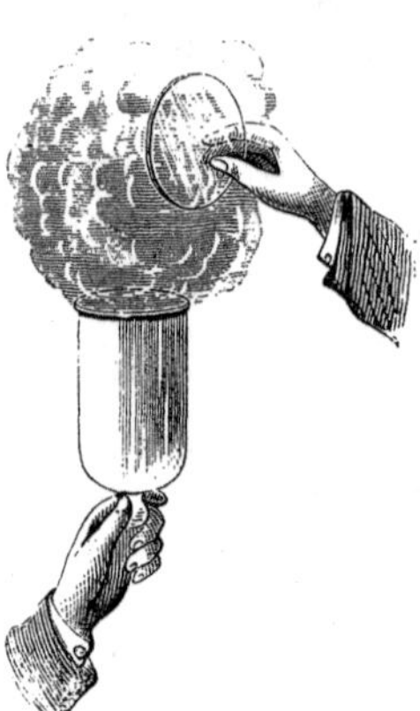
Fig. 53. N_2O_2 and air.

On removing the cover of a jar of nitrogen di-oxide, Fig. 52, and immersing in the gas a long slip of blue litmus paper, the lower end of the paper in the pure gas will be unaffected, while the upper end in contact with the red fumes produced by union with the oxygen of the air, will be turned red. A somewhat large bell-glass filled with this gas, gives voluminous clouds of the brown-red vapors when its cover is removed, as shown in Fig. 53. This experiment demonstrates the existence of free oxygen in the air.

HYPONITROUS OXIDE.—*Formula* N_2O. *Molecular weight* 44. *Molecular volume* 2. *Density* 22. *One liter weight* 1·97 *grams (22 criths).*

235. History.—Hyponitrous oxide was discovered by Priestley in 1776. It was more minutely examined by Davy in 1809, who discovered its exhilarating property. It was first used as an anæsthetic by Wells in 1845.

236. Preparation.—Hyponitrous oxide may be prepared by reducing nitric acid with zinc or stannous chloride; but it is generally obtained by decomposing ammonium nitrate by heat, according to the equation:—

$$(NH_4)NO_3 = N_2O + (H_2O)_2$$

The gas, being heavier than air, may be collected by displacement. It may also be collected over warm water. The apparatus used, is shown in Fig. 54.

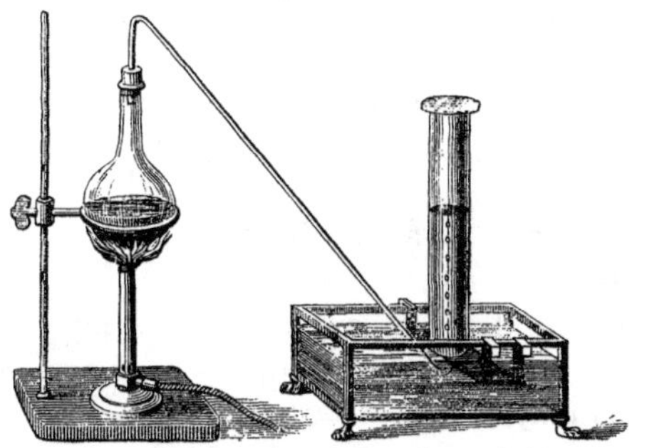

Fig. 54. Preparation of Hyponitrous Oxide.

237. Properties.—Hyponitrous oxide, sometimes called nitrous oxide, is a colorless gas, inodorous, but with a distinctly sweet taste. It is one half heavier than air, its specific gravity being 1·527. Subjected to a pressure of thirty-two atmospheres at 0°, it is condensed to a colorless mobile liquid of specific gravity 0·9, which freezes at −101°. The liquid also freezes by its own evaporation when allowed to escape into the open air, producing a snow-like mass, which, mixed with carbon disulphide and placed in a vacuum, produces the lowest temperature yet obtained,—140°. The gas is quite soluble in water, one hundred volumes dissolving seventy-eight volumes at 15°. It is more soluble in alcohol and in alkaline solutions.

Burning bodies have their combustion accelerated in hyponitrous oxide. A candle, having a spark upon the wick, is relighted in it, much as in oxygen. Phosphorus and sulphur burn in it with great splendor. The gas is decomposed and its oxygen unites with the combustible. When breathed in moderate quantity, it exerts a marked exhilarative action upon the system, and hence has been called laughing gas. Of late years, it has come extensively into use as an anæsthetic agent, the inhalation being continued for a longer time. It was with this gas that the property of anæsthesia was discovered; a discovery everywhere acknowledged as the crowning surgical discovery of the present century.

Its composition may be ascertained by passing electric sparks through it, when it separates into two volumes of nitrogen and one of oxygen; or by exploding it with an equal volume of hydrogen, when its own volume of nitrogen only is left.

§ 2. Phosphorus.

Symbol P. *Atomic weight* 31. *Molecular weight* 124. *Molecular volume* 2. *Equivalence* I, III *and* V. *Density* 62. *One liter of phosphorus-vapor weighs* 5·55 *grams (62 criths).*

238. History.—Phosphorus was discovered in 1669, by Brandt, by igniting evaporated urine in closed vessels. One hundred years later, in 1769, Gahn and Scheele discovered it in bones and in 1775 proposed a method of preparing it from them.

239. Occurrence.—Phosphorus does not occur free in nature. It exists in combination, in the minerals apatite, pyromorphite, wagnerite, etc., which are calcium, lead, and magnesium phosphates, respectively. Vast deposits of calcium phosphate occur on many of the Caribbean islands, and near Charleston, S. C. The bones of animals contain calcium phosphate, and this as well as other phosphates are present in their tissues, being derived mostly from the seeds of plants.

240. Preparation.—Phosphorus is prepared by acting upon burned bones with sulphuric acid, leaching off the resulting liquid, evaporating it to dryness, and distilling the residue with charcoal. The earthy matter of bones consists of calcium phosphate $Ca''_3(PO_4)_2$. By treating this with sulphuric acid, acid calcium phosphate results, as follows :—

$$\underset{\text{Calcium phosphate.}}{Ca''_3(PO_4)_2} + \underset{\text{Sulphuric acid.}}{(H_2SO_4)_2} = \underset{\text{Calcium sulphate.}}{(CaSO_4)_2} + \underset{\text{Hydro-calcium phosphate.}}{H_4Ca''(PO_4)_2}$$

This is leached off and by evaporation to dryness, converted into calcium meta-phosphate, thus :—

$$H_4Ca''(PO_4)_2 = Ca''(PO_3)_2 + (H_2O)_2;$$

and the calcium meta-phosphate, distilled with charcoal

Fig. 55. Preparation of Phosphorus.

gives phosphorus and calcium phosphate again, according to the equation :—

$$(Ca''(PO_3)_2)_3 + C_{10} = Ca''_3(PO_4)_2 + (CO)_{10} + P_4.$$

Practically, the bones previously burned and ground very fine, are mixed with two-thirds of their weight of strong sulphuric acid diluted with eighteen or twenty parts of water, well stirred, and allowed to stand for twelve hours. The clear liquid is then strained from the deposited calcium sulphate or gypsum, evaporated in a pan to a syrupy consistence, mixed with one-fifth its weight of charcoal powder, and heated to low redness. The dry mass is then placed in earthen retorts with long necks, and these are raised gradually to bright redness in the furnace shown in Fig. 55, when the phosphorus distills over and condenses in the receivers. Theoretically, the bone ash should yield eleven per cent of phosphorus, but practically only eight per cent is obtained.

The crude phosphorus is purified generally by melting it under water and agitating it with a mixture of potassium di-chromate and sulphuric acid. The impurities

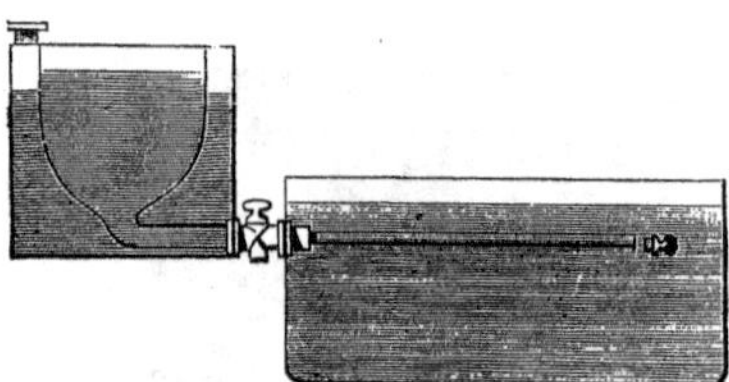

Fig. 56. Casting Phosphorus in Sticks.

are oxidized, and the pure liquid phosphorus remains colorless and transparent at the bottom of the vessel. It is then ladled into a conical vessel surrounded with warm water, shown in Fig. 56, from the bottom of which a tube passes, through a stopcock, into another tube laid horizontally in a vessel containing cold water, which tube can be closed by a plug. On opening the cock the tube fills with melted phosphorus, which, in the cold water soon solidifies, and may be withdrawn as a solid stick by removing the plug. In this form it is brought into commerce.

241. Properties.—Phosphorus is capable of existing in two markedly different allotropic states. Prepared as above, α phosphorus is a colorless, transparent, wax-like solid, having a specific gravity of 1·83. It melts at 44° to a colorless liquid and boils at 290°, yielding a colorless vapor of specific gravity 4·355. It crystallizes from its solution in carbon disulphide, in the form of the regular dodecahedron, Fig. 57, 1. It is not soluble in water, but dissolves easily in carbon disulphide, in phosphorous chloride, in alcohol, in ether, and in certain volatile and fixed oils. In the air it oxidizes readily and hence must be kept under water. Owing to this slow combustion it is luminous in the dark; though a trace of naphtha or oil of turpentine in the air, prevents this phenomenon. It is violently poisonous, and kills by depriving the blood of oxygen. Oil of turpentine is the best antidote. Heated to 50° in the air it takes fire and burns vividly, forming phosphoric oxide.

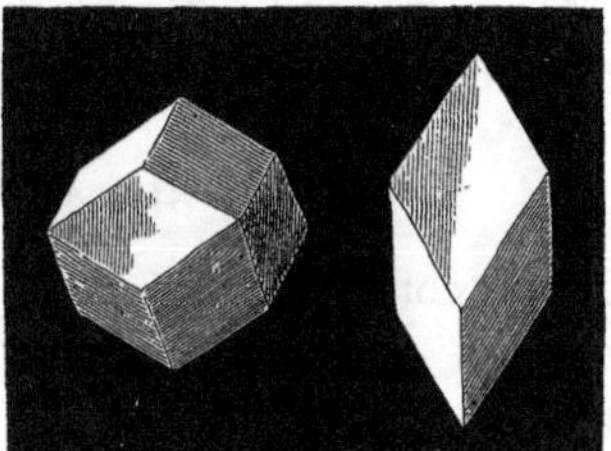

1. Fig. 57. 2.
Phosphorus Crystals.

In 1848, Schrötter discovered that by heating ordinary phosphorus to 250° in a gas which had no action upon it, it was converted into a chocolate-red powder—β phosphorus—possessing properties entirely different from those previously exhibited. Its specific gravity is 2·14. At 250°–260° it melts, being re-converted into α. It is insoluble in the ordinary solvents of phosphorus, but it may be dissolved in metallic lead by heating in a sealed tube with this metal. On cooling it crystallizes out in acute rhombohedral crystals, Fig. 57, 2, having a metallic luster, and a specific gravity of 2·34. It has no odor, does not oxidize readily in the air and is not poisonous. It does not take fire until heated to 260°.

242. Uses.—Phosphorus is extensively used in the manufacture of friction matches. For this purpose the α variety is generally employed; though on account of the frightful disease of the jaw which it causes in the workmen, the β or red variety is much to be preferred. It is used also in medicine, and as a rat-poison.

PHOSPHORUS AND HYDROGEN.

Three compounds of phosphorus and hydrogen are known: a gaseous compound H_3P, a liquid one H_4P_2, and a solid one H_2P_4.

HYDROGEN PHOSPHIDE OR PHOSPHINE.—*Formula* H_3P. *Molecular weight* 34. *Molecular volume* 2. *Density* 17. *One liter weighs* 1·52 *grams* (17 *criths*).

243. History.—Hydrogen phosphide was discovered in 1783 by Gengembre; but its analogy with ammonia was first established by Heinrich Rose, in 1832.

244. Preparation.—It may be prepared by the action of water or acids upon phosphides:—

$$Ca_3P_2 + (HCl)_6 = (CaCl_2)_3 + (H_3P)_2;$$

by the decomposition of so-called hypophosphorous acid:—

$$(H(PO_2H_2))_2 = H_3PO_4 + H_3P;$$

and by boiling phosphorus with a solution of potassium hydrate :—

$$(HKO)_3 + P_4 + (H_2O)_3 = (K(PO_2H_2))_3 + H_3P.$$

245. Properties.—Phosphine is a colorless gas, with a nauseous garlic-like odor. It is sparingly soluble in water, is condensable to a liquid, and is neutral in its reaction. Its specific gravity is 1·134. It takes fire readily at 100°, burning with a brilliant flame. It unites directly with hydriodic acid forming phosphonium iodide, analogous to ammonium iodide formed from ammonia in the same way.

EXPERIMENT.—Hydrogen phosphide may be conveniently prepared by the apparatus shown in Fig. 58. The retort is one-third filled with moderately concentrated solution of potassium hydrate, a few pieces of phosphorus are dropped in, and by means of the tube

Fig. 58. Preparation of Hydrogen Phosphide.

passing through the tubulure, the air is displaced by a current of pure hydrogen. The beak of the retort is prolonged by a glass tube which dips beneath the surface of the water in the porcelain dish, thus cutting off contact with the air. On heating the contents of the retort to boiling, hydrogen gas first escapes, but soon bubbles of phosphine make their appearance, each one of which as it bursts takes fire spontaneously—owing to a small quantity of the liquid phosphide dissolved in it—and forms a beautiful ring of white smoke which rotates on its circular axis as it ascends. If the air of the room be still, several of these rings will follow each other to the ceiling. When the experiment is concluded, the hydrogen may be again passed until the gas is no longer spontaneously inflammable.

OXIDES AND ACIDS OF PHOSPHORUS.

The normal oxides and acids of phosphorus form a parallel series with those of nitrogen. The known members of the series are the following :—

Oxides.		*Acids.*	
Hypophosphorous oxide	P'_2O (?)	—	—
Phosphorous oxide	P'''_2O_3	—	—
Phosphoric oxide	$P^V{}_2O_5$	Hypophosphorous acid	$H(P^VO_2H_2)$
		Phosphorous acid	$H_2(P^VO_3H)$
		Phosphoric acid	$H_3(P^VO_4)$
		Meta-phosphoric acid	$H(P^VO_3)$

PHOSPHORIC OXIDE.—*Formula* P_2O_5. *Molecular weight* 142.

246. Preparation.—Phosphoric oxide is always the product of the rapid combustion of phosphorus in the air or in oxygen. The reaction is synthetic, thus :—

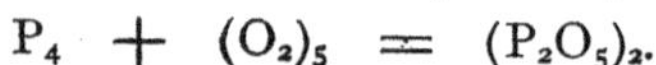

$$P_4 + (O_2)_5 = (P_2O_5)_2.$$

Fig. 59. Formation of Phosphoric oxide

EXPERIMENT.—Place a fragment of carefully dried phosphorus in a small cup on a stand, in the middle of a dining-plate; ignite it by a hot wire, and cover it with a large bell-glass, as shown in Fig. 59. White fumes will fill the jar, gradually aggregate together and fall into the plate, resembling in appearance a miniature snow-storm.

247. Properties.—Phosphoric oxide is a snow-white amorphous powder, which is fusible at a red heat and easily volatilized. It rapidly attracts moisture from the air, and adheres together in flocks. When plunged into water it hisses like a hot iron, and then dissolves, forming phosphoric acid.

TRI-HYDROGEN PHOSPHATE OR PHOSPHORIC ACID.—*Formula* H_3PO_4. *Molecular weight* 98.

248. Preparation.—Phosphoric acid is obtained by the action of boiling water upon phosphoric oxide:—

$$P_2O_5 + (H_2O)_3 = (H_3PO_4)_2;$$

by the oxidation of phosphorus by nitric acid; or by the decomposition of phosphates by sulphuric acid:—

$$\underset{\text{Lead phosphate.}}{Pb''_3(PO_4)_2} + \underset{\text{Sulphuric acid}}{(H_2(SO_4))_3} = \underset{\text{Lead sulphate.}}{(Pb''(SO_4))_3} + \underset{\text{Phosphoric acid}}{(H_3PO_4)_2}.$$

Commercially, an impure acid is prepared by treating bone-ash—calcium phosphate—with sulphuric acid.

249. Properties.—As thus prepared, phosphoric acid is a syrupy liquid, which by spontaneous evaporation over sulphuric acid, gives hard, transparent, prismatic crystals, which deliquesce in the air. Their solution is intensely acid; it does not coagulate albumin, nor precipitate barium chloride. It throws down from ammoniacal solutions of magnesium sulphate, a white crystalline precipitate of ammonio-magnesium phosphate, sometimes called triple phosphate. It gives with silver nitrate, when neutralized by ammonia, yellow silver phosphate. On heating its aqueous solution to 213°, it gives diphosphoric acid; and on raising the temperature to redness, meta-phosphoric acid is produced.

This form of phosphoric acid, being tribasic, is capable of forming acid, normal, and double salts. The following are examples of each :—

Acid Salts.

Di-hydro-sodium phosphate	H_2NaPO_4
Hydro-di-sodium phosphate	HNa_2PO_4
Hydro-calcium phosphate	$HCa''PO_4$

Normal Salts.

Potassium phosphate	K_3PO_4
Barium phosphate	$Ba''_3(PO_4)_2$
Bismuth phosphate	$Bi'''PO_4$

Double Salts.

Ammonio-magnesium phosphate	$(NH_4)Mg''(PO_4)$
Potassio-barium phosphate	$KBa''(PO_4)$

MONO-HYDROGEN PHOSPHATE OR META-PHOSPHORIC ACID.—*Formula* HPO_3. *Molecular weight* 80.

250. History and Preparation.—In 1833 Graham showed that ordinary phosphoric acid lost water on being heated to redness, and on cooling, became a transparent ice-like solid, the so-called glacial phosphoric acid :—

$$H_3PO_4 - H_2O = HPO_3.$$

The same acid is produced by dissolving phosphoric oxide in cold water :—

$$P_2O_5 + H_2O = (HPO_3)_2.$$

Meta-phosphates are produced by igniting acid phosphates which have two hydrogen atoms :—

$$H_2NaPO_4 - H_2O = NaPO_3,$$

or which have two atoms of volatile base :—

$$\underset{\text{Hydro-ammonio-sodium phosphate.}}{H(NH_4)NaPO_4} = \underset{\text{Ammonium hydrate.}}{H(NH_4)O} + \underset{\text{Sodium meta-phosphate.}}{NaPO_3}.$$

By decomposing meta-phosphates, the acid is obtained.

251. Properties.—Meta-phosphoric acid is a hard, transparent, colorless, glassy mass, not crystallizable, and very soluble in water, forming a strongly acid solution, which gradually takes up water and forms tri-hydrogen phosphate. It coagulates albumin and gives a white precipitate with silver nitrate. It is monobasic, and forms but one class of salts. It is distinguished by a remarkable tendency to produce polymeric forms, called di-, tri-, tetra-, and hexa-meta-phosphates, respectively.

The two phosphoric acids now described are the mono- and di-meta forms of ortho-phosphoric acid, H_5PO_5, being only more distinctly marked examples under a general law. Certain ortho-phosphates, as the mineral libethenite, hydro-di-cupric ortho-phosphate HCu''_2PO_5, and di-hydro-ammonio-magnesium phosphate $H_2(NH_4)Mg''PO_5$, are well known bodies.

252. Di-phosphoric or Pyro-phosphoric Acid.—*Formula* $H_4P_2O_7$. *Molecular weight* 178.

In 1826, Clark discovered a variety of phosphoric acid intermediate between the two forms already described, produced by heating a solution of the tri-basic acid to 213°, and which for this reason, he called pyrophosphoric acid. Two molecules of the tri-hydrogen phosphate together lose one of water:—

$$(H_3PO_4)_2 \quad - \quad H_2O \quad = \quad H_4P_2O_7.$$

Di-phosphates are produced by igniting a phosphate which has one atom of volatile base:—

$(HNa_2PO_4)_2$	$-$	H_2O	$=$	$Na_4P_2O_7$.
Hydro-di-sodium phosphate.		*Water.*		*Sodium di-phosphate.*

It occurs generally in solution, but may be obtained by evaporation at 213° as a soft glass, or in semi-crystalline masses. Its solution is strongly acid, does not coagulate albumin, and precipitates silver nitrate white. Being tetrabasic, di-phosphoric acid forms a large series of acid, normal and double salts. It bears the same relation to tribasic phosphoric acid, that di-sulphuric acid bears to dibasic sulphuric acid. On boiling its solution, it takes up a molecule of water and becomes tri-hydrogen phosphate; on igniting it, it loses one, becoming mono-hydrogen phosphate.

253. Aldehydic Phosphoric Acids.—Two other acids of pentad phosphorus are known, which, as they resemble the aldehydes of organic chemistry, may be called aldehydic acids. These are commonly known as phosphorous and hypo-phosphorous acids. Phosphorous acid has the formula $H_2(PO_3H)$, is dibasic, and forms so-called phosphites. Hypo-phosphorous acid, $H(PO_2H_2)$ is monobasic, and is obtained by decomposing barium hypo-phosphite with sulphuric acid. Hypo-phosphites are formed by boiling phosphorus in alkaline solutions.

PHOSPHOROUS OXIDE. — *Formula* P_2O_3. *Molecular weight* 110.

254. Preparation and Properties.—Phosphorous oxide is produced by the imperfect combustion of phosphorus in dry air. For this purpose the phosphorus is placed in a somewhat narrow tube, Fig. 60, drawn to a point at

Fig. 60. Production of Phosphorous oxide.

one end and at the other connected with an aspirator, by means of which air may be drawn through the tube. On heating the tube, the phosphorus takes fire and a bulky amorphous deposit of phosphorous oxide collects beyond it in the tube. The white flakes are readily volatile and have an alliaceous odor. They are deliquescent and dissolve in water with a hissing noise, forming an acid solution. The rational constitution of this compound is unknown; but it can hardly be normal, since it does not give the normal acid by the action of water.

No acid of triad phosphorus is known.

255. Hypophosphorous Oxide.—By covering fragments of phosphorus with a layer of phosphorous chloride, and exposing the whole to the air, Leverrier obtained a canary-yellow substance, soluble in water, and decomposing when boiled, into phosphoric acid and a flocculent substance having the composition P_4O.

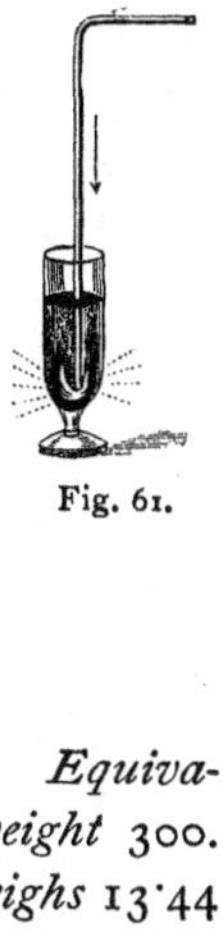

EXPERIMENT.—By melting a piece of phosphorus under warm water, Fig. 61, and then passing into it a current of oxygen from a gas-holder, the phosphorus will take fire and burn brilliantly beneath the water. The phosphoric acid produced dissolves in the liquid; but beside this a red-brown powder is formed, which was formerly regarded as hypophosphorous oxide, but which is now believed to be only impure amorphous phosphorus.

Fig. 61.

§ 3. ARSENIC AND ANTIMONY.

ARSENIC.—*Symbol* As. *Atomic weight* 75. *Equivalence* III *and* V. *Density* 150. *Molecular weight* 300. *Molecular volume* 2. 1 *liter of arsenic vapor weighs* 13·44 *grams (*150 *criths).*

256. History.—Two sulphides of arsenic occur native, one red, the other yellow. The former is mentioned by Aristotle under the name σανδἄϱάκη, and the latter by Dioscorides under the name *ἀρσενικόν*, from which the name arsenic is derived. The metal was first obtained by Schroeder in 1694, and was more minutely examined by Brandt in 1733.

257. Occurrence. — Arsenic occurs somewhat abundantly in nature, both free and combined with other metals as iron, copper, cobalt, nickel, etc. The most abundant sources of it are the iron arsenides leuco- and arseno-pyrite. The yellow sulphide called orpiment, and the red sulphide, or realgar, also occur native.

258. Preparation.—From mispickel or arsenical pyrites arsenic is obtained by heating it in earthen tubes or retorts. The arsenic being volatile, sublimes and condenses in the cooler portions of the retort, toward its mouth. In certain districts arsenic is obtained by reducing its oxide with charcoal; this method gives it in a purer form.

259. Properties.—Arsenic is a dark steel-gray brittle solid with a metallic luster, and a specific graxity of 5·6 to 5·9. It occurs in two, perhaps three, allotropic modifications. Beside the steel-gray variety α, which crystallizes in rhombohedrons, and has the above specific gravity, there is an amorphous black vitreous variety β, of specific gravity 4·71, which at 360° passes into α, with considerable evolution of heat. Arsenic is volatile at 180°; its vapor is orange-yellow and has a peculiar odor resembling garlic. In the air, it gradually oxidizes at common temperatures, and at a red heat it burns with a bluish-white flame, producing arsenous oxide. Arsenic and all its compounds are active poisons. In the arts it is used in pyrotechny, in the manufacture of shot, and as a fly-powder under the name of cobalt.

ARSENIC AND HYDROGEN.

HYDROGEN ARSENIDE OR ARSINE.—*Formula* H_3As. *Molecular weight* 78. *Molecular volume* 2. *One liter weighs* 3·48 *grams* (39 *criths*).

260. History.—Hydrogen arsenide was discovered by Scheele in 1755.

261. Preparation.—It is always prepared by the action of pure zinc upon sulphuric or hydrochloric acid containing arsenic, or of these acids in the pure state upon zinc containing arsenic :—

As_2Zn_3	+	$(HCl)_6$	=	$(Zn\,Cl_2)_3$	+	$(H_3As)_2$
Zinc arsenide		*Hydrogen chloride.*		*Zinc chloride.*		*Hydrogen arsenide.*

262. Properties. — Arsine is a colorless gas, with an alliaceous odor, and a specific gravity of 2·7. Cooled to −40° it becomes a liquid. It is soluble in five times its volume of water. It takes fire easily in the air, burning with a bluish-white flame, evolving white fumes of arsenous oxide. If a cold surface of porcelain be held in this flame, metallic arsenic is deposited upon it as a black stain or tache. Hydrogen arsenide is easily decomposed when the tube through which it is passing is heated to redness, a black mirror-like ring of metallic arsenic being formed just beyond the heated spot. The gas is also decomposed when passed into a solution of silver nitrate, forming arsenous acid and precipitating metallic silver.

EXPERIMENT.—Marsh's test for arsenic depends upon the production of arsine whenever arsenic is present in any soluble form in a solution in which hydrogen is being evolved. The form of Marsh's apparatus employed by the author is represented in Fig. 62. It consists of a three-necked bottle, through the middle tubulure of which a funnel tube passes for the supply of liquid, while one of the side openings has a siphon tube for withdrawing the exhausted acid, and the other a delivery tube carrying a bulb filled with cotton, to retain impurities mechanically carried over with the gas. Next to this bottle is a jar containing potassium hydrate and calcium chloride to purify the gas, which then passes through a long tube of hard glass drawn out at intervals. Pure zinc in frag-

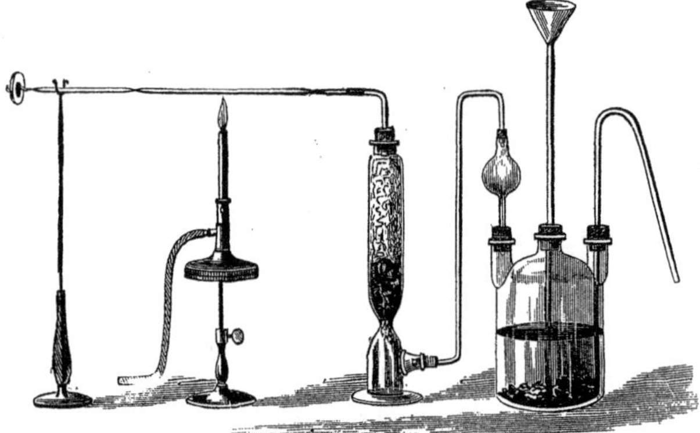

Fig. 62. Marsh's Arsenic Apparatus.

ments is first put into the three-necked bottle, and then pure sulphuric acid, previously diluted with three parts of water and cooled, is added until this bottle is about one-third full. After allowing sufficient time for the air to be expelled from the apparatus, the narrow tube is heated to dull redness by the gas flame. If no dark deposit appears beyond the flame in fifteen minutes, the materials may be considered pure. The liquid suspected to contain arsenic, —which must be perfectly free from all organic matter—is now added through the funnel-tube. If arsenic be present, the flame of hydrogen burning at the end of the tube will, often in a few seconds, change color, becoming whitish, and will deposit a dark brown metallic spot on the porcelain crucible-cover pressed down upon it. If the tube be again heated, the arsine will be decomposed and the arsenic be deposited as a black metallic ring. According to Wormley $\frac{1}{50000}$ of a grain of arsenous oxide in one hundred grain measures of the solution may be detected by this test.

Hydrogen arsenide is one of the most active poisons known. It should therefore be experimented upon with the greatest care.

OXIDES AND ACIDS OF ARSENIC.

The oxides of arsenic, with their corresponding acids, are the following:—

Oxides.		*Acids.*	
Arsenous oxide	As_2O_3	Arsenous acid	$HAsO_2$
Arsenic oxide	As_2O_5	Arsenic acid	H_3AsO_4

ARSENIC OXIDE.—*Formula* As_2O_5. *Molecular weight* 230.

263. Preparation and Properties. — Arsenic oxide, obtained by heating arsenic acid to dull redness, is an opaque, white, amorphous, deliquescent mass, which fuses at a bright red heat, and decomposes into arsenous oxide and oxygen. By solution in water it forms arsenic acid.

ARSENIC ACID. — *Formula* H_3AsO_4. *Molecular weight* 142.

264. Preparation and Properties.—Arsenic acid is produced by oxidizing arsenous oxide or acid by nitric acid, and evaporating to a syrup. On standing, long rhomboidal laminæ separate, which contain water of crystallization and are deliquescent. At 100° this water is expelled and needle-shaped crystals of the acid H_3AsO_4 are produced. Its aqueous solution is strongly acid. On heating arsenic acid to 150°, di- or pyro-arsenic acid, $H_4As_2O_7$, results ; and at 200°, another molecule of water is lost and meta-arsenic acid $HAsO_3$, is obtained.

Salts, corresponding to each of these acids, have been produced. Arsenic acid and its salts are poisonous, though less so than arsenous compounds.

ARSENOUS OXIDE. — *Formula* As_2O_3. *Molecular weight* 198. *Molecular volume* 1 (*anomalous*). *Density* 198.

265. Occurrence and Preparation.—Arsenous oxide occurs native as the mineral arsenolite. It is prepared by roasting arsenical ores with free access of air, and collecting the vapors in partitioned chambers. The fine dust thus obtained is purified by re-sublimation.

266. Properties.—Arsenous oxide exists in two, probably polymeric, modifications. When condensed at a temperature of 400°, it forms a transparent vitreous mass α, of specific gravity 3·738. When deposited slowly at temperatures slightly less elevated, this variety crystallizes in right rhombic prisms. The second modification β, is obtained by condensing the vapor at 200°, in brilliant transparent octahedral crystals, of specific gravity 3·689. The same octahedral form is obtained on cooling a saturated aqueous solution. The vitreous or α variety passes gradually at ordinary temperatures, rapidly at 100°, into β, forming a white opaque mass resembling porcelain. When the vitreous variety is dissolved to saturation in hot hydrochloric acid and left to cool slowly, it crystallizes in octahedrons, the formation of each crystal being accompanied with a flash of light. Arsenous oxide volatilizes at 218°, yielding a vapor whose density is 198 instead of 99. This would indicate that the vitreous modification, which is formed at high temperatures, has the molecular formula As_4O_6, double that of the octahedral modification.

Both varieties are soluble in water, in hydrochloric acid and in alkaline solutions. Arsenous oxide is a most energetic poison, one or two decigrams being sufficient to destroy life.

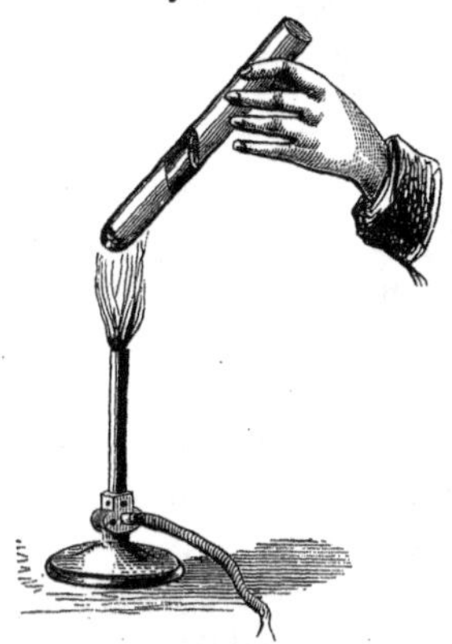

Fig. 63. Testing for Arsenic.

EXPERIMENTS.—Arsenous oxide may be readily recognized by its weight, by its volatility, and by its crystalline form. If a few milligrams be placed in a small open tube, Fig. 63, and heated over a gas-flame—the upper part of the tube being slightly warmed—the oxide will be volatilized and the vapor will condense on the tube in a ring of brilliant octahedral crystals like those in Fig. 64. If in another similar tube a mixture of arsenous oxide and a little charcoal be placed and heated, a black ring of metallic arsenic will be deposited, as shown in the figure. No other substance but arsenic gives these appearances.

Fig. 64. Crystals of Arsenous oxide.

The best antidote for arsenic is freshly prepared ferric or magnesic hydrate.

ARSENOUS ACID. — *Formula* H_3AsO_3. *Molecular weight* 126.

267. Preparation and Properties.—When arsenous oxide is dissolved in water, an acid, styptic liquid is obtained, which cannot be evaporated without decomposition. The arsenites in general are more stable; they are both ortho-arsenites and meta-arsenites. Potassium arsenite is the chief ingredient in Fowler's solution, used in medicine, and copper arsenite in Scheele's green, used as a pigment.

ANTIMONY.—*Symbol* Sb. *Atomic weight* 122. *Equivalence* III *and* V. *Density* 244 *(?)*. *Molecular weight* 488 *(?)*. *Molecular volume* 2. 1 *liter of antimony-vapor weighs* 21·86 *grams (*244 *criths) (?)*.

268. History and Occurrence.—Antimony was first prepared by Basil Valentine toward the end of the fifteenth century. It occurs in nature both free and in combination. The most abundant source of it is the sulphide, known as stibnite; but it exists in combination with oxygen in the minerals valentinite, senarmontite, and cervantite; with silver, in dyscrasite; and with silver and sulphur in pyrargyrite and miargyrite.

269. Preparation and Properties. — Commercial antimony is generally produced by acting upon the melted native sulphide with iron, producing ferrous sulphide and free antimony, according to the equation:—

$$Sb_2S_3 + Fe_3 = (FeS)_3 + Sb_2.$$

It is also obtained by roasting the sulphide, and then reducing the oxide thus obtained, with charcoal. By fusion with sodium carbonate and a small quantity of antimonous sulphide, the metal may be obtained pure.

Antimony is a brilliant bluish-white brittle metal, of specific gravity 6·7. It crystallizes in rhombohedrons, isomorphous with arsenic and red phosphorus. A curious allotropic variety is obtained by electrolysis, having a specific gravity of 5·8, and passing into the ordinary condition when heated or struck, with the evolution of great heat. Antimony melts at 450° and at a white heat, may be distilled. It tarnishes scarcely at all in the air, but takes fire at a red heat, producing antimonous oxide. It is strongly attacked by chlorine, forming antimonous and antimonic chlorides, $SbCl_3$ and $SbCl_5$. Antimony is used largely in the arts as a constituent of type-metal.

COMPOUNDS OF ANTIMONY.

270. Hydrogen Antimonide or Stibine, H_3Sb.—Whenever an antimony compound is present in a solution from which hydrogen is being evolved, an inodorous gas escapes mixed with the hydrogen, causing it to burn with a bluish-white flame. This gas is stibine. It is analogous to arsine, and is similarly decomposed by heat; but the metallic deposit of antimony is easily distinguished from that of arsenic by its darker color, its smoky appearance, its less volatility, its insolubility in hypochlorites, and its solubility in ammonium sulphide.

271. Antimonous and Antimonic Oxides and Acids.—Antimonic oxide Sb_2O_5, is a tasteless, yellowish, insoluble powder, of specific gravity 6·6, obtained by heating antimonic acid. Antimonic acid is obtained by oxidizing antimony by nitric acid or by treating antimonic chloride with water. Both the ortho-acid, H_5SbO_5, and the di-meta acid $HSbO_3$, have been obtained; and salts of di-antimonic acid $H_4Sb_2O_7$ are also known.

Antimonous oxide, Sb_2O_3, occurs native, in two different crystalline forms, as senarmontite and valentinite. It is the product of the combustion of antimony in the air, and is then crystalline; by pouring antimonous chloride into a boiling solution of sodium carbonate it is obtained as a dirty-white powder, which becomes yellow when heated. Antimonous acid, $HSbO_2$, is a feeble acid, forming easily decomposable salts.

272. Antimonous and Antimonic Sulphides and Sulph-acids.—Antimonic sulphide Sb_2S_5, is obtained as a yellowish-red powder by the action of hydrogen sulphide upon a solution of the chloride in tartaric acid, or by acidifying a solution of an alkaline sulph-antimonate. It unites readily with sulphides of positive elements, forming sulph-antimonates, having the general formula M_3SbS_4; of which sodium sulph-antimonate Na_3SbS_4,9 aq.—sometimes called Schlippe's salt—is an example.

Antimonous sulphide, Sb_2S_3, exists native as stibnite, as above stated. It has a steel-gray color, a specific gravity of 4·5, and a strong metallic luster. It crystallizes in ortho-rhombic prisms. It is thrown down by hydrogen sulphide from antimonous solutions as a bright orange-red precipitate, which contains water. On fusion it becomes steel-gray. The sulph-antimonites, which it forms by union with positive sulphides, are largely represented among minerals: pyrargyrite Ag'_3SbS_3, and boulangerite $Pb''_3(SbS_3)_2$, being ortho-sulph-antimonites; miargyrite $Ag'SbS_2$, and berthierite $Fe''(SbS_2)_2$, being meta-sulph-antimonites; and jamesonite $Pb''_2Sb_2S_5$, and brongniardite $(PbAg)''_2Sb_2S_5$, being di- or para-sulph-antimonites.

§ 4. Bismuth.

Symbol Bi. *Atomic weight* 210. *Equivalence* III *and* V. *Specific gravity of solid* 9·83. *Fuses at* 264°.

273. History.—Bismuth was first distinctly recognized

by Basil Valentine in the 15th century. Agricola in 1529, calls it Bisemutum, and Paracelsus mentions it as Wisemat. It was for a long time confounded with other metals, especially with lead, tin, and antimony. Pott in 1739, first described its characteristic reactions.

274. Occurrence and Preparation.—Bismuth occurs in the metallic state in veins traversing gneiss, clay-slate, and other crystalline rocks, principally in Saxony and Bohemia. It occurs also as oxide, forming the mineral bismite; as sulphide, or bismuthinite; as sulpho-telluride or tetradymite; and as carbonate or bismutite. It is prepared on the large scale in the arts from the native bismuth, by placing this, mixed with the rocky gangue, in iron tubes, slightly inclined, which are heated in a furnace. The bismuth melts and flows out at the lower ends of the tubes into suitable vessels, from which it is ladled into moulds. The bismuth of commerce contains arsenic, iron, and other metals, from which it may be freed by fusion with potassium nitrate, by which they are oxidized. Chemically pure bismuth may be obtained by reducing the basic nitrate by charcoal.

275. Properties.—Bismuth is a hard, brittle, brilliant reddish-white metal. It is powerfully dia-magnetic and has a strong tendency to crystallize when cooled from fusion; by melting a considerable quantity of it, allowing it to cool until a crust forms on the surface, piercing this, and pouring out the metal which still remains fluid, crystals of great size and beauty may be obtained. They are rhombohedrons, though on account of the large interfacial angle, 87°40′, they have often been mistaken for cubes, in which this angle is 90°. Owing to a slight superficial oxidation, these crystals, as usually obtained, are beautifully iridescent. Bismuth has a specific gravity of 9·83 ; it melts at 264°, and expands one-thirty-second of its bulk on solidifying. It is unaltered in dry air, but is tarnished in presence of moisture. Strongly heated, it takes fire, burning with a bluish-white flame, and forming bismuthous oxide. Chlorine and nitric acid attack it readily ; but hydrochloric and sulphuric acids, when cold, have no action upon it.

Bismuth is used in the arts for forming alloys. Rose's fusible metal is composed of one part of lead, one of tin, and two of bismuth ; it melts at 94°. Lipowitz's fusible metal contains three parts of cadmium, four of tin, eight of lead, and fifteen of bismuth ; it melts at 60°. An alloy of lead and bismuth is used in the so-called permanent metallic pencils.

COMPOUNDS OF BISMUTH.

276. Bismuthous Chloride, $BiCl_3$.—This chloride may be formed by the direct action of chlorine upon bismuth. It is a white, granular, deliquescent substance, readily fusible and volatile. By contact with water it is decomposed, forming bismuthyl chloride, (BiO)Cl, sometimes called bismuth oxychloride.

277. Bismuthic Oxide, Bi_2O_5.—Bismuthic oxide is obtained by heating its hydrate to 130°. It is a brown powder, which gives up a part of its oxygen readily. Its hydrate, called bismuthic acid, $HBiO_3$, may be prepared by oxidizing bismuthous hydrate, suspended in water, by a current of chlorine. It is a bright red powder which loses oxygen a little above 100°. Its salts are called bismuthates ; potassium bismuthate, $KBiO_3$, and bismuthous bismuthate, $Bi'''Bi^{V}O_4$, are examples.

278. Bismuthous Oxide, Bi_2O_3.—This oxide occurs native as bismite. It may be formed by burning the metal in air, or by igniting the hydrate, carbonate, or nitrate. It is a pale-yellow powder, of specific gravity 8·2, which melts at a red heat, and is insoluble in water. It has been obtained in ortho-rhombic prisms. Bismuthous hydrate, H_3BiO_3, is obtained by treating a bismuthous salt, as the chloride, for example, with potassium hydrate. A white, flocculent precipitate is thrown down, which on drying loses water, and becomes an amorphous white powder, $HBiO_2$. With strong bases this body acts like an acid, sodium bismuthite, $NaBiO_2$, being one of its salts. But with strong acids, on the other hand, this hydrate is strongly basic, the nitrate $Bi(NO_3)_3$, the sulphate, $Bi_2(SO_4)_3$, the carbonate, $Bi_2(CO_3)_3$, and the phosphate, $BiPO_4$, being well-known compounds. By pouring the nitrate into a large quantity of water, the mono-meta-nitrate, $BiNO_4$, is produced.

Bismuthous sulphide, Bi_2S_3, occurs native as bismuthinite. It is obtained as a black precipitate, on adding hydrogen sulphide to solutions of bismuth. It forms with metallic sulphides, salts called sulpho-bismuthites, some of which, as that of lead or kobellite, $Pb''_3(BiS_3)_2$, and those of copper, called emplectite, $Cu''(BiS_2)_2$, and wittichenite, $Cu''_3(BiS_3)_2$, occur native.

§ 5. Relations of the Nitrogen Group.

279. The members of the nitrogen group have close relations, both physically and chemically, with each other. From nitrogen to bismuth they increase in density and in atomic weight, while their chemical activity decreases in the same order. The last four are iso-dimorphous; that is, they all have two crystal-forms, which are the same for each substance. A comparison of their compounds will show how closely they are allied chemically:—

TRIAD COMPOUNDS.

	Hydrides.	*Chlorides.*	*Oxides.*	*Sulphides.*
N	H_3N	NCl_3	N_2O_3	——
P	H_3P	PCl_3	P_2O_3	P_2S_3
As	H_3As	$AsCl_3$	As_2O_3	As_2S_3
Sb	H_3Sb	$SbCl_3$	Sb_2O_3	Sb_2S_3
Bi	——	$BiCl_3$	Bi_2O_3	Bi_2S_3

PENTAD COMPOUNDS.

	Chlorides.	*Oxides.*	*Sulphides.*
N	——	N_2O_5	——
P	PCl_5	P_2O_5	P_2S_5
As	$AsCl_5$	As_2O_5	As_2S_5
Sb	$SbCl_5$	Sb_2O_5	Sb_2S_5
Bi	——	Bi_2O_5	——

EXERCISES.

§ 1.

1. In what ways may nitrogen be prepared?
2. A given volume of oxygen weighs thirty grams; what is the weight of the same volume of nitrogen?
3. What weight of oxygen does a cubic meter of air contain?
4. What are the proofs that air is merely a mixture?
5. What volume of air is required to yield six kilograms of oxygen?
6. At what temperature has air the density of H at 0°?
7. What weight of nitrogen do ten grams of ammonia contain?
8. A liter of water is to be saturated with ammonia-gas at 0°; how many grams of $(NH_4)Cl$ and of CaO must be used in the process?

9. What volume of H_3N at 15° will a kilogram of NH_4Cl yield?

10. 250 c. c. ammonia decomposed by electric sparks gives what volume of mixed gases? If 200 c. c. of oxygen be added and exploded, what will be the composition of the remaining gas?

11. What volume of NH_3 will neutralize a liter of HCl?

12. How much nitric acid may be obtained from a kilogram KNO_3 by the laboratory process? How much by the commercial?

13. To neutralize ten grams MgO, requires how many c. c. of nitric acid of sp. gr. 1·42? (Acid contains 70 per cent of HNO_3.)

14. Write the graphic formula of nitrogen tetr-oxide.

15. A kilogram of copper gives what volume of N_2O_2?

16. One liter of N_2O requires what volume of oxygen to make N_2O_4?

17. How many c. c. of O and of N in one liter of nitrogen di-oxide?

18. Fifteen grams ammonium nitrate yield what volume of N_2O?

19. What volume of H will one gram of N_2O burn? One liter?

20. One c. c. liquefied N_2O_2 requires what weight of $(NH_4)NO_3$?

§ 2.

21. Give the chemical stages in making phosphorus.

22. One hundred kilograms of bone-ash, 80 per cent calcium phosphate, yields how many kilograms of phosphorus?

23. How do common and allotropic phosphorus differ?

24. One liter of phosphine contains what volume of P vapor?

25. What volume of air is required to burn ten grams of P?

26. If dissolved in boiling water, how much phosphoric acid would the above product yield?

§ 3.

27. Give the preparation and properties of metallic arsenic.

28. Describe Marsh's test. What is the limit of its delicacy?

29. At a certain temperature 100 liters of hydrogen weigh four grams; what will this volume of As vapor weigh at the same temperature?

30. H_3As contains what volume of H? How much air is required to burn it?

31. Calculate the percentage of antimony in antimonous sulphide.

32. How is stibine distinguished from arsine?

33. Give the formulas of the common antimony compounds?

CHAPTER FIFTH.

BORON.

Symbol B. *Atomic weight* 11. *Equivalence* III. *Molecular weight* 22 (?). *Molecular volume* 2. *Sp. gr.* 2·68.

280. History and Occurrence.—Under the Arabic name buraq, corrupted into borax, a salt obtained from certain lakes in Thibet, and containing boron as an essential component, has long been imported into Europe. From this, in 1702, Homberg obtained boric oxide; and from boric oxide, Davy, in 1807, by the aid of electricity, and Gay-Lussac and Thenard in 1808, by chemical means, obtained pure boron. It was first obtained crystallized by Wöhler and Deville in 1856. The mineral sassolite is boric acid, H_3BO_3; and borax, boracite, and larderellite, are native borates of sodium, magnesium and ammonium respectively.

281. Preparation and Properties.— Amorphous boron may be prepared by the action of sodium upon potassium fluoborate, thus :—

$$(KBF_4)_2 + Na_6 = (KF)_2 + (NaF)_6 + B_2.$$

In this form, boron is a soft, chocolate-brown powder, slightly soluble in water, and fusible at the heat of the oxy-hydrogen flame. It may be obtained crystallized by dissolving it in melted aluminum, allowing the mass to cool, and removing the aluminum by hydrochloric acid. Short quadratic octahedrons are left undissolved, which vary from honey-yellow to garnet-red in color, have a specific gravity of 2·63, and are nearly as hard, as lustrous, and as highly refractive as the diamond itself. These crystals are infusible, and are combustible with difficulty even in oxygen. They are not attacked by melted niter. The so-called graphitoidal boron is a compound of boron and aluminum.

282. Boric Oxide. B_2O_3.—Boric oxide is formed whenever boron burns in the air or in oxygen. It is usually obtained by igniting its hydrate, boric acid. A viscid mass is left, which solidifies to a colorless, brittle, transparent glass, of specific gravity 1·83. It unites directly with positive oxides to form borates, expelling the more volatile negative oxides with which they are combined.

283. Boric Acid, H_3BO_3.—Boric acid occurs free in nature, especially in volcanic districts, as in Tuscany, where it issues, with steam and gaseous matters, from fissures in the earth, into natural or artificial ponds or lagoons, the water of which soon becomes charged with the acid. This water is then evaporated and the acid crystallized out. Boric acid may be prepared from sodium borate or borax, by dissolving three parts of it in twelve of boiling water, adding one part of sulphuric acid, and allowing the whole to cool. The boric acid separates in white, crystalline scales, of specific gravity 1·48, soluble in two and a half parts of water at 18°, and freely soluble in alcohol. Its aqueous solution reddens litmus paper, and turns turmeric paper brown. Its solution in alcohol burns with a green flame. This is the normal or ortho-boric acid. By heating it to 120°, it loses one molecule of water and yields meta-boric acid, HBO_2. Boric acid shows a strong tendency to condensation, thus forming multiple salts. Borax, or s o d i u m t e t r a b o r a t e, $Na_2B_4O_7$, is an example ; it is found native, in the waters of certain lakes in Thibet and California. It is used largely as a flux in working metals.

CHAPTER SIXTH.

NEGATIVE TETRADS.

§ 1. CARBON

Symbol C. *Atomic weight* 12. *Equivalence* II *and* IV. *Density* 12 *(?). Molecular weight* 24 *(?). Molecular volume* 2. 1 *liter of carbon-vapor weighs* 1·075 *grams (*12 *criths) (?).*

284. Occurrence.—Carbon occurs native in two allotropic forms, known as the diamond and as graphite. Also more or less impure, in the various forms of mineral coal. Combined with hydrogen, it occurs in bitumen and petroleum ; with oxygen, it exists in the air ; and with oxygen and calcium, it forms limestone, a very abundant rock, of which twelve per cent is carbon. It is an essential constituent too, of all animal and vegetable tissues.

285. Properties.—I. AS DIAMOND.—The form of carbon known as the diamond is a brilliant, transparent, and generally colorless solid, having a specific gravity of 3·5 and crystallizing in forms belonging to the isometric system, Fig. 65, the faces being often rounded. It does not conduct heat or electricity, and has a very high refractive and dispersive power. It is the hardest form of matter known. Lavoisier established its composition in 1776, by burning it in oxygen.

The diamond occurs generally in the form of small rounded pebbles in alluvial detritus, produced by the disintegration of ancient rocks, from which it is obtained

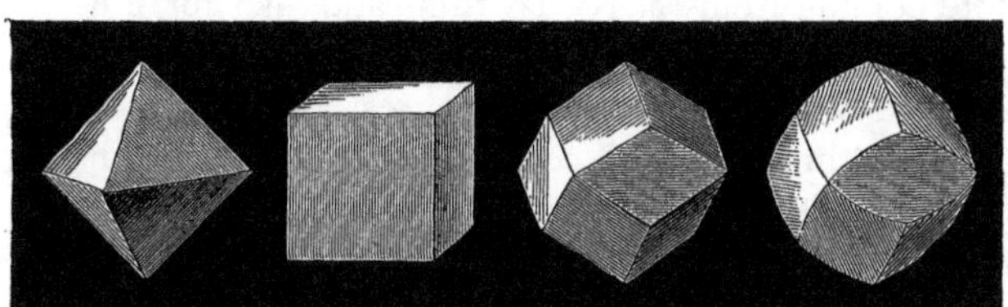

Fig. 65. Crystalline forms of Diamond.

by washing. The chief localities are India, Borneo, and Brazil; though a few diamonds have been found in Georgia and North Carolina in this country. It is cut for a gem into forms having a relation to its directions of cleavage—known as the brilliant, the rose, and the table—by means of diamond dust.

II. As Graphite. — The second form of carbon, known as graphite, is a friable, leaden-gray solid, unctuous to the touch, of specific gravity 2 to 2·2, and crystallizes in hexagonal plates, Fig. 66. It has a semi-metallic luster, conducts heat and electricity readily, and is combustible with difficulty, leaving generally a few percents of ash. It is soluble in melted iron, from which it crystallizes on cooling.

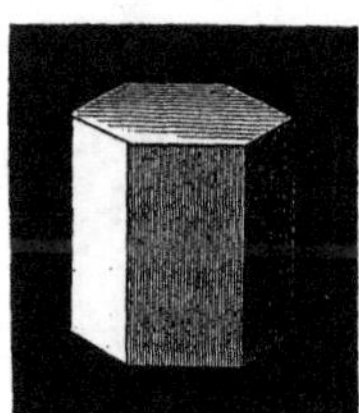

Fig. 66. Form of Graphite Crystal.

Graphite occurs both foliated and massive, in metamorphic rocks in England, Siberia and Ceylon, and in various localities in the United States; as Sturbridge, Mass., Ticonderoga, N. Y., Brandon, Vt., Wake, N. C., etc. It is purified by Brodie's process, by treating it with potassium chlorate and nitric acid ; and is, after drying, condensed to a solid block by hydrostatic pressure. It is largely used for making pencils,—the name graphite coming from γραφω, I write—and also for crucibles.

III. As Mineral Coal.—The purest variety of carbon in the form of mineral coal, is that known as anthracite. It is an amorphous, hard, lustrous, black solid, difficultly combustible, and consisting of from 80 to 94 per cent carbon. Its specific gravity varies from 1·3 to 1·7. From this there is a regular gradation through cannel and bituminous coals of all varieties, to lignite or brown coal, which in some cases, is scarcely altered wood. All coal is derived from primitive vegetation, changed and consolidated by heat and pressure. Anthracite coals are found where the strata have been most disturbed, bituminous, where they remain nearly or quite horizontal ; while brown coal or lignite is more recent in age, being generally tertiary.

Other varieties of more or less pure carbon, are : gas-carbon or plumbagine, deposited in the cast-iron retorts in which coal-gas is made ; a hard, compact, mammillated variety, being almost metallic in appearance, and conducting both heat and electricity. Specific gravity 1·76. Vegetable coal or charcoal, prepared on the large scale by burning wood in heaps as shown in Fig. 67, and for special purposes, in iron cylinders ; a bluish-black, porous substance, retaining minutely the form of the original wood, and having a specific gravity of 1·7. Animal coal or bone-black, obtained by igniting bones in close vessels ; also a black porous mass, containing 90 per cent of calcium phosphate. And lamp-black, or soot, prepared by collecting the matters condensed from the smoke of highly carbonized bodies such as pitch and tar, in stone chambers; a soft, finely-divided, and very impure form of carbon, used in printing-ink and as a pigment.

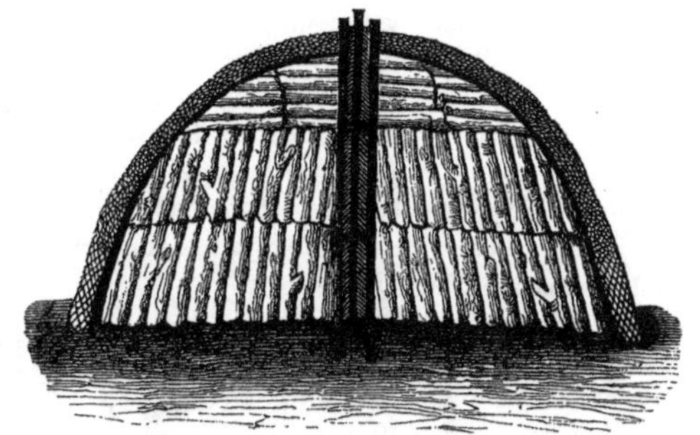

Fig. 67. Interior of Charcoal Heap.

In all its forms, carbon is infusible and non-volatile. As vegetable and animal coal, owing to its very great porosity—a single cubic centimeter of box-wood charcoal exposing a surface of near half a square meter—carbon exhibits a remarkable power of absorbing gases. Thus, box-wood charcoal will absorb 90 volumes of ammonia gas, and that made from the shell of the cocoanut, 171 volumes. From this fact, i. e., that charcoal carries within it a condensed mass of the oxygen of the air in which it cooled, it acts as an energetic disinfectant by oxidizing foul vapors. For the same reason, animal charcoal is largely used as a decolorizing agent, particularly in refining sugar. It has also been proposed for filling respirators, which are placed over the mouth to absorb noxious gases.

EXPERIMENTS.—The powerful absorption of ammonia gas by

Fig. 68. Absorption of Gases by Charcoal.

charcoal may be shown by filling a cylinder over mercury with the dry gas, Fig. 68, and then introducing into it a piece of charcoal, previously heated to redness in sand, and allowed to cool away from the air. The ammonia will be rapidly absorbed by the charcoal and the mercury will rise in the tube.

To show the decolorizing property of charcoal, place in four flasks, dilute solutions of indigo, cochineal, iodide of starch, and potassium permanganate. Agitate each with recently ignited bone-black, and throw each upon a separate filter, Fig. 69. The liquors as they run from the funnels will be colorless. If beer or ale be thus treated, it will lose not only its color, but also its bitter taste.

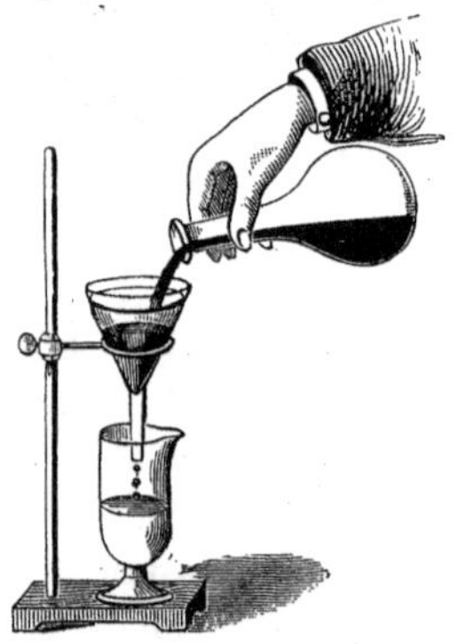
Fig. 69. Decolorizing power of Charcoal.

Carbon is apparently unalterable in the air at ordinary temperatures. Charred piles driven by the Britons to prevent Julius Cæsar from crossing the Thames, and wheat charred nearly 2,000 years ago at Herculaneum, are yet unchanged. When heated in the air, it burns, forming the di-oxide. Heated with sulphur or hydrogen, it also unites directly with these bodies, forming carbon disulphide and acetylene.

CARBON AND HYDROGEN.

286. Hydrocarbons.—The compounds of carbon and hydrogen—called hydrocarbons—are very numerous, and are usually classified in series. Most of them are more satisfactorily considered in Organic Chemistry. Only three of them will therefore be described here. These are, hydrogen carbide H_4C, hydrogen di-carbide H_4C_2, and di-hydrogen di-carbide H_2C_2.

HYDROGEN CARBIDE, OR METHANE.—*Formula* H_4C. *Molecular weight* 16. *Molecular volume* 2. *Density* 8. *One liter weighs* 0·716 *grams (*8 *criths).*

287. Occurrence.—Methane occurs free in nature, being produced somewhat abundantly by the decomposition of vegetable matter confined under water. It constitutes 75 per cent of the gas which rises when the bottom of a pond covered with vegetable matter is stirred with a stick, whence it is often called marsh-gas. It also escapes from seams in coal mines and constitutes the fire-damp of miners. It often occurs largely in the vicinity of salt-wells, as in Kanawha, West Va. The town of Fredonia, N. Y., is wholly or partially lighted by it.

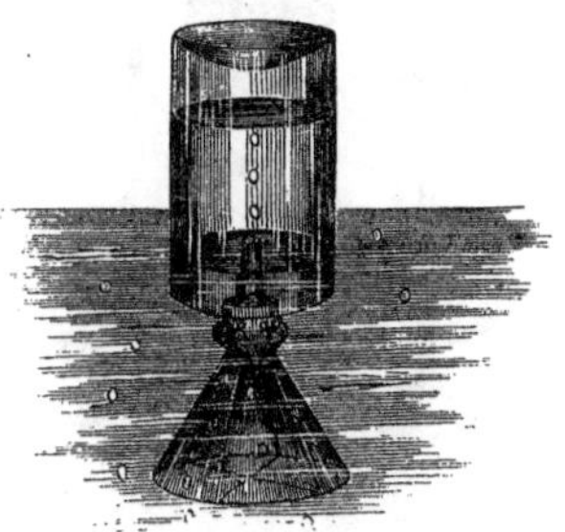

Fig. 70. Collection of Marsh-gas.

288. Preparation.—Marsh-gas may be obtained by filling a bottle with water, attaching a funnel to its mouth by a string, as shown in Fig. 70, and inverting it in a pond so as to catch the bubbles which rise on stirring the mud at the bottom. By agitating it with a little lime-water, it may be purified for experiment.

Methane may be also procured by heating potassium acetate in presence of a strong base, usually potassium or sodium hydrate. The reaction which takes place is as follows:—

$$CO\left\{\begin{matrix}CH_3\\ONa\end{matrix}\right. + \left.\begin{matrix}Na\\H\end{matrix}\right\}O = CO\left\{\begin{matrix}ONa\\ONa\end{matrix}\right. + H_4C.$$

Sodium acetate. *Sodium hydrate.* *Sodium carbonate.* *Methane.*

EXPERIMENT.—For the preparation of methane, two parts crystallized sodium acetate, two parts sodium hydrate, and three parts

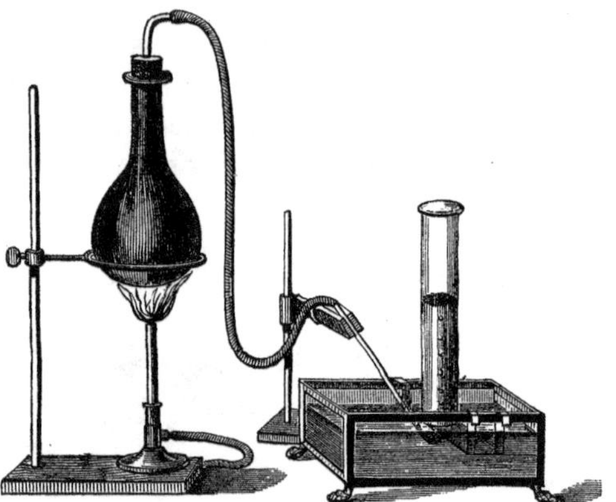

Fig. 71. Preparation of Hydrogen Carbide.

powdered quick-lime, are intimately mixed together and heated in a thin copper or iron flask to bright redness, Fig. 71. The lime is used to prevent the fusion of the mass. The gas is rapidly evolved and may be collected over water.

289. Properties.—Hydrogen carbide is a colorless, odorless, and tasteless gas, but slightly soluble in water, and not condensable to a liquid by cold or pressure. It is the lightest gas next to hydrogen, its specific gravity being 0·5576. It is combustible, burning in the air with a pale, faintly-luminous flame. By passing electric sparks through the gas, it is decomposed, and yields twice its volume of hydrogen. It forms an explosive mixture with air, and is the cause of the serious coal-mine explosions which sometimes happen in coal districts. By the action of chlorine, its hydrogen is gradually replaced by this element, forming successively the compounds CH_3Cl, CH_2Cl_2, $CHCl_3$, and CCl_4.

Methane constitutes the first of a homologous series of hydrocarbons known as the marsh-gas series. They are all saturated bodies, having the general formula C_nH_{2n+2}; i. e., they contain twice as many atoms of hydrogen as of carbon, plus two. They constitute the essential portion of the various native petroleums.

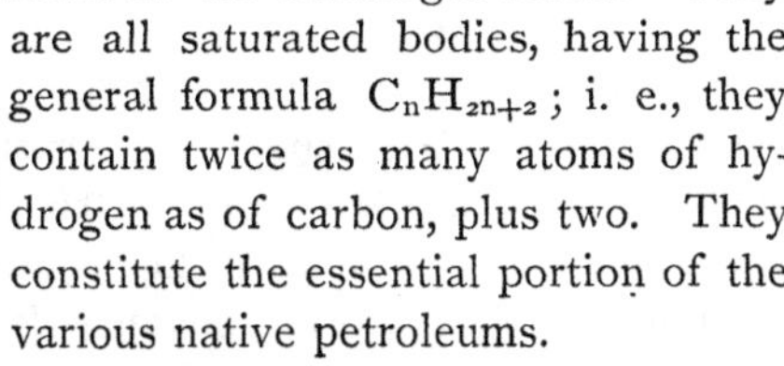
Fig. 72. Combustibility of Marsh-gas.

EXPERIMENT.—The levity and inflammability of this gas may be shown, as in the case of hydrogen, by introducing a lighted taper into a jar of it, held mouth downward. The gas will burn at the mouth of the jar, and the candle-flame, as it passes up into it, will be extinguished.

HYDROGEN DI-CARBIDE OR ETHYLENE. — *Formula* H_4C_2. *Molecular weight* 28. *Molecular volume* 2. *Density* 14. *One liter weighs* 1·25 *grams (*14 *criths).*

290. History.—Ethylene was discovered in 1796 by four Dutch chemists, Deiman, Paets van Troostwyk, Bondt, and Lauwerenburgh. It has been found in small quantity among the gases of coal-mines.

291. Preparation.—It is usually prepared by the action of sulphuric acid upon alcohol, according to the reaction:—

$$C_2H_6O = H_4C_2 + H_2O.$$

This equation exhibits only the final result; the chemical change itself is evidently more complicated.

EXPERIMENTS.—To prepare ethylene, one volume of alcohol and four volumes of sulphuric acid are mixed in a flask with sand to a thick paste, and gently heated. The sand is added to prevent frothing toward the end. The gas may be purified by passing it through milk of lime to remove sulphurous oxide, and through strong sulphuric acid to retain the vapors of ether and alcohol.

Mitscherlich's continuous process consists in passing the vapor of 80 per cent alcohol through boiling dilute sulphuric acid—three parts of water to ten of acid.

292. Properties.—Hydrogen di-carbide is a colorless, irrespirable gas, having usually an etherial odor. Its specific gravity is 0·978. Under strong pressure at $-110°$, it condenses to a limpid liquid, which at $1°$ has a vapor-tension of 42½ atmospheres. It is soluble in about eight times its volume of water. It is readily combustible, burning in the air with a brilliant white flame, evolving much smoke. Mixed with three volumes of oxygen, it explodes violently on the approach of a flame. It is decomposed by the electric spark, the carbon being deposited, and twice its volume of hydrogen remaining. It unites directly with an equal volume of chlorine, forming an oily liquid, ethylene chloride. From this fact, the gas received from its discoverers the name "olefiant gas;" and the liquid was called the "oil of the Dutch chemists."

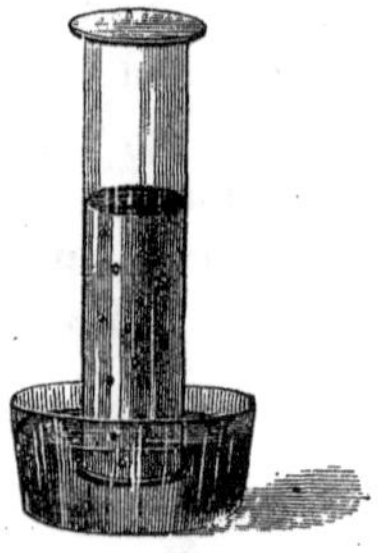

EXPERIMENT.—To show the direct union of ethylene and chlorine, fill a glass cylinder half full of chlorine over the water-cistern, and then add rapidly an equal volume of olefiant gas. The gaseous mixture will at once begin to diminish in volume, Fig. 73, oily drops will collect upon the walls of the vessel, and, sinking through the fluid, form a liquid layer upon the bottom of the containing cistern. By pouring off the water and agitating with sodium carbonate solution, the

Fig. 73. Formation of Ethylene Chloride.

ethylene chloride may be purified and its agreeable, chloroform-like odor obtained.

Di-hydrogen Di-carbide or Acetylene.—*Formula* H_2C_2. *Molecular weight* 26. *Molecular volume* 2. *Density* 13. *One liter weighs* 1·16 *grams (* 13 *criths).*

293. History and Preparation.—Acetylene was discovered by E. Davy in 1836, and studied by Berthelot in 1860. It may be prepared by the direct union of its constituents. When the carbon points terminating the electrodes of a powerful voltaic battery are brought together in an atmosphere of hydrogen, the carbon and hydrogen combine at the elevated temperature produced, and form acetylene. It is also a product of the action of heat upon substances rich in carbon and hydrogen, and is always formed in the imperfect combustion of hydrocarbon compounds.

294. Properties.—Acetylene is a colorless gas, having a peculiar and disagreeable odor, and not condensable to a liquid. It has a specific gravity of 0·92, is quite soluble in water, and burns with a bright but very smoky flame. It is readily absorbed by ammoniacal cuprous chloride, forming a red precipitate of cuprous acetylide, which is explosive. This explosive body is sometimes formed in brass gas-pipes by the action upon them of the acetylene in coal-gas, and has been the cause of fatal accidents.

ILLUMINATING-GAS.

295. History.—The production of a combustible gas from coal was first observed by Clayton in 1664; but it was not until 1792 that Murdoch made gas-illumination a practical success. In 1798, he lighted, in this way, Boulton and Watt's works at Soho, near Birmingham. The streets of London were first lighted with gas in 1812. Gas was introduced into Paris in 1815.

296. Preparation.—Illuminating-gas is ordinarily prepared by distilling bituminous coal at a high temperature; though various other substances, such as oil, rosin, wood, and petroleum, have also been employed for its preparation. The complete apparatus for the manufacture, purification, and collection of coal-gas, is represented in Fig. 74. The coal is placed in semi-cylindrical iron retorts, C, set in a furnace shown upon the right, their mouths being closed by heavy plates. Usually five retorts are heated by the same fire, forming what is technically called a "bench." The products of the distillation pass from the retort through a tube near its mouth, up into a larger horizontal tube, B, called the hydraulic main, where the tar and a portion of the water are condensed to liquids; the gas then passes on, first through a series of vertical pipes, D, and then through the coke-box or 'scrubber' O, by which it is still further cooled and the condensable vapors separated. It then enters the purifier, M, a large metallic box containing, on shelves for the purpose, either dry slaked lime or what is preferable, ferric hydrate, either alone or mixed with lime and sawdust. From this it issues, freed from most of its impurities, particularly sulphur-compounds and carbonic di-oxide, and is collected in the adjoining gasometer, G, for distribution.

EXPERIMENT.—The manufacture of gas may be illustrated upon the lecture-table by the apparatus shown in Fig. 75. The coal is placed in the retort upon the right, which is then heated by the gas-burner. The water and volatile liquid products condense in the

Fig. 74 Preparation of Coal-gas.

receiver, while the gas passes on to the first U-tube, in one limb of which a piece of red litmus-paper is placed,—to detect ammonia,—and in the other a strip of paper moistened with a solution of lead acetate,—to detect hydrogen sulphide,—and then to the second,

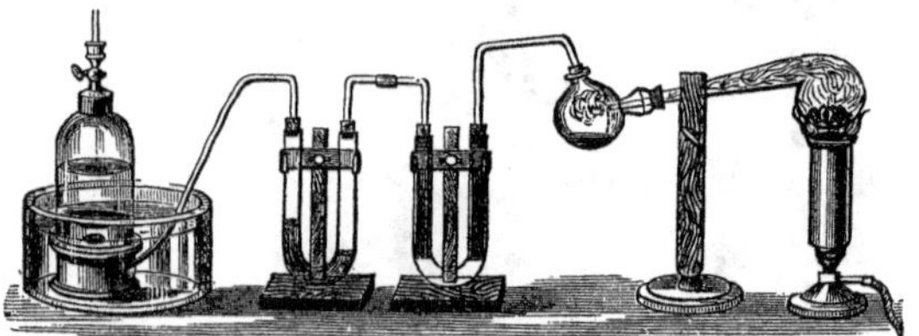

Fig. 75. Distillation of Coal.

the bend of which contains lime-water—which indicates, by becoming milky, the presence of carbonic di-oxide,—and is finally collected over water in the capped receiver. By depressing this receiver in the water, and opening the cock, the gas as it issues from the jet may be lighted.

297. Composition and Properties.—Coal-gas is a mixture of several gaseous products which vary according to the quality of coal used, the temperature at which it is distilled, etc., but which consist essentially of hydrogen and methane (marsh-gas) mixed with variable proportions of olefiant gas, acetylene, carbonous and carbonic oxides, butylene, nitrogen, oxygen, and hydrogen sulphide. The amount of gas obtained varies, with different coals, from 8,000 to 15,000 cubic feet to the ton. Coal-gas has a specific gravity varying from 0·65 to 0·34. The illuminating power of a gas is determined by an instrument called a photometer, in which the amount of light given by the gas, burning from a jet at the rate of five cubic feet per hour, is compared with that emitted by a standard candle burning 120 grains of spermaceti in the same time. A gas may rise in illuminating power to 25 or 30 candles; but the average supplied in our cities is about 16 candles.

The collateral products of the coal-gas manufacture are in general, two; the ammoniacal liquors, and the gas-tar. The former consists of the condensed water, holding in solution the ammonia produced from the nitrogenous matters in the coal; the latter is a complex substance, containing in its lighter portions certain volatile liquids as benzol and toluol; and certain volatile alkaline bases as aniline and chinoline; and in its heavier, certain phenols as phenol proper (carbolic acid) and cresol, and certain solid hydrocarbons as naphthalin and anthracene.

CARBON AND OXYGEN.

CARBONIC DI-OXIDE. — *Formula* CO_2. *Molecular weight* 44. *Molecular volume* 2. *Density* 22. *One liter weighs* 1·97 *grams (22 criths).*

298. History.—Carbonic di-oxide was the first gas distinguished from air. It was noticed as a distinct substance by Paracelsus in 1520; and soon after, Van Helmont obtained it from limestone,—whence he called it chalky air,—and noticed its production in the fermentation of sugar and in the burning of charcoal, and its occurrence naturally. Black showed in 1757 that alkalies absorbed it and that its compounds effervesced with acids. Lavoisier in 1775 determined its composition synthetically, by burning carbon in oxygen.

299. Occurrence.—Carbonic di-oxide exists in the air to the extent of about ·04 per cent, being produced by combustion, fermentation, respiration, etc. In volcanic districts it is more abundant, the large quantity of it often collected in low places frequently proving fatal to life. Combined with calcium oxide or lime, it exists in enormous quantity in limestone.

300. Preparation.—Carbonic di-oxide may be prepared by direct synthesis. It is always the product of the combustion of carbon in air or in oxygen :—

$$C + O_2 = CO_2.$$

It is generally obtained however, by the action of an acid upon some carbonate, as that of sodium or of calcium:

$$CaCO_3 + (HNO_3)_2 = Ca''(NO_3)_2 + H_2O + CO_2.$$

It is produced in large quantities in burning limestone for the production of quicklime :—

$$CaCO_3 = CaO + CO_2.$$

Fig. 76. Effervescence of Carbonates with Acids.

EXPERIMENTS. — All carbonates effervesce with acids from the setting free of carbonic di-oxide gas. If some fragments of marble be placed in the test-glass, Fig. 76, and some hydrochloric acid be added, a brisk effervescence will take place from the escape of this gas. In order to collect it, the experiment may be repeated in the two-necked bottle, Fig. 77, the acid being poured in through the funnel tube upon the marble, previously covered with water. The gas may be collected over water, or by displacement, as it is heavier than air.

301. Properties.—Carbonic di-oxide is a colorless gas with a slightly pungent odor and acid taste. It is heavier than air, its specific gravity being 1·524. At the ordinary temperature and pressure, water dissolves its own volume of this gas. As the pressure increases, the temperature remaining the same, another volume is absorbed for each atmosphere added ; but as the gas is condensed in the same ratio, according to Marriotte's law, it follows that the volume of the gas dissolved by water is the same at all pressures, while the weight is directly proportional to the pressure. Subjected to a pressure of 38·5 atmosphere at 0°, it condenses to a colorless limpid liquid, of specific gravity 0·83 at 0°.

Fig. 77. Preparation of Carbonic Di-oxide.

When a fine stream of the liquefied gas is allowed to escape into the air, the rapid evaporation of one portion freezes another, producing a snow-white flocculent mass, which may be formed into balls like snow, and which disappears with great slowness. Moistened with ether and placed in the vacuum of an air-pump, a temperature of −110° may be obtained. Carbonic dioxide extinguishes the combustion of burning bodies placed in it and is fatal to animal life, though less actively than was formerly supposed. Diluted largely with air it exerts a narcotic action, and has been proposed as an anæsthetic. Fatal effects have resulted from entering wells, fermenting vats, and other places in which this gas has accumulated. Before going into such a place, a lighted candle should be lowered into it; if it is extinguished, the place is unsafe.

EXPERIMENTS.—Carbonic di-oxide may be condensed to a liquid by using the apparatus either of Thilorier, in which the gas is liquefied by its own pressure; or of Natterer, in which the

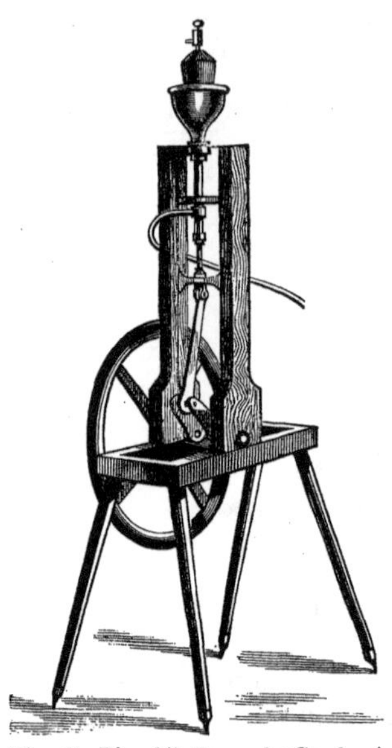
Fig. 78. Bianchi's Pump for Condensing Gases.

pressure is produced mechanically. Fig. 78 represents Bianchi's modification of Natterer's condensing pump. The piston is solid and is worked by a crank and fly-wheel as shown in the figure. The receiver at the top of the pump-barrel is made of heavy cannon-metal, and has a tight valve below, and a screw-plug above, by which the liquefied gas may be drawn off. The tube leading to the pump serves to convey the dried carbonic di-oxide; as the pump is worked, the pressure in the receiver increases until it reaches 38·5 atmospheres — the receiver being surrounded with ice—when each additional stroke of the pump liquefies the gas which it forces into it. In this way, half a kilogram of liquid carbonic di-oxide may be obtained in a short time. The receiver is then removed from the pump and inverted, and the gas allowed to escape into a peculiarly constructed cylindrical box of metal, which in the course of a few seconds, is filled with the solid carbonic di-oxide snow. On moistening some of it with ether in a wooden tray, and immersing in the mixture thermometer-bulbs full of mercury, bullets of frozen mercury are easily obtained.

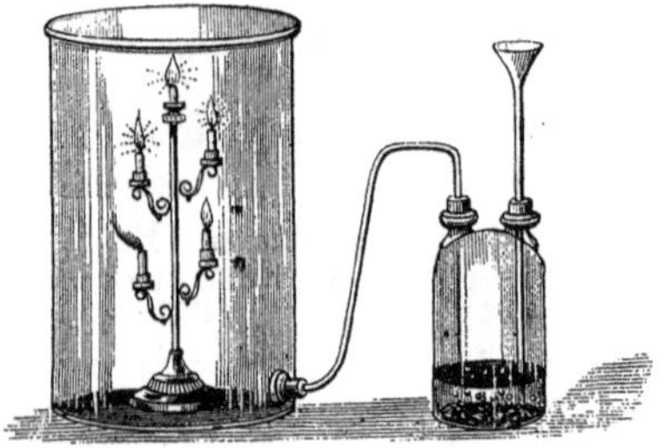
Fig. 79. Candles extinguished by Carbonic Di-oxide

From the ease with which carbon di-oxide gas can be prepared, it is used to illustrate many of the properties of gases. Its weight may be shown by balancing carefully a large beaker or a thin pasteboard box, on one of the scale-pans of a large balance, and pouring into it a second large beakerful of the gas. The scale-pan will at once descend. A candle lighted, and placed in the bottom of a beaker, is extinguished on pouring some of the gas upon it. If a stand carrying a series of lighted candles be placed in a large jar, Fig. 79, and carbonic di-oxide gas be introduced at the bottom, the candles will be successively extinguished as the gas rises around them. The accumulation of this gas in wells, and its removal therefrom by buckets, may be shown by using a tall jar, Fig. 80, to represent the well, and a glass bucket attached to a wire to draw up the gas. If a lighted candle be near, the gas may be poured from the bucket upon the flame so as to put it out, Fig. 81. This is the only gas which extinguishes flame and renders lime-water milky. If a little clear lime-water be poured into a jar of the gas it becomes at once turbid from the production of calcium carbonate.

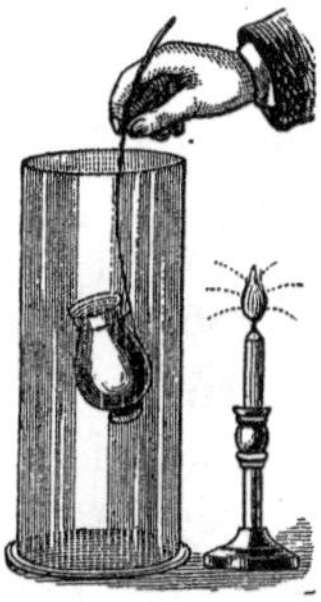

Fig. 80. Drawing the gas from a well.

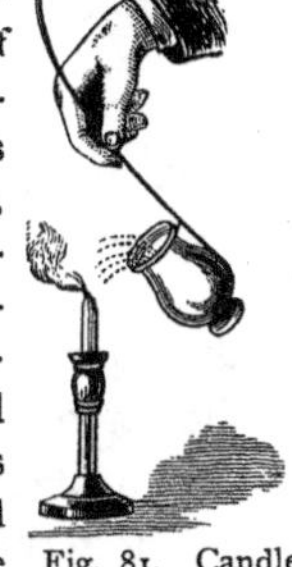

Fig. 81. Candle extinguished by CO_2.

302. Carbonic Acid. — *Formula* H_2CO_3. *Molecular weight* 62. Carbonic acid is produced by the solution of carbonic di-oxide in water. It has not been obtained free from water, since it readily decomposes into water and carbonic di-oxide again on slightly raising the temperature. The solution in water is a distinctly acid liquid having the pungent odor and agreeable acid taste so well known in the so-called soda-water, which is made simply by condensing carbonic di-oxide gas into water.

The salts of carbonic acid are called carbonates. Both ortho-carbonates and meta-carbonates, having respectively the general formulas M'_4CO_4 and M'_2CO_3, are known.

Carbonous Oxide, or Carbonyl. — *Formula* CO. *Molecular weight* 28. *Molecular volume* 2. *Density* 14. *One liter weighs* 1·25 *grams* (14 *criths*).

303. History. — Carbonous oxide was discovered by Lassone in 1776, and also by Priestley in 1783. Its true nature was determined by Woodhouse in 1800.

304. Preparation.—Carbonous oxide may be prepared (1) by the incomplete combustion of carbon, as when charcoal and blacksmith's scales—ferroso-ferric oxide—are heated together :—

$$Fe_3O_4 + C_4 = Fe_3 + (CO)_4 ;$$

(2) by the abstraction of oxygen from carbonic di-oxide, as when this gas is passed over heated charcoal :—

$$CO_2 + C = (CO)_2 ;$$

(3) by heating oxalic acid with strong sulphuric acid :—

$$\underset{\text{Oxalic acid.}}{H_2C_2O_4} = \underset{\text{Water.}}{H_2O} + \underset{\text{Carbonic di-oxide.}}{CO_2} + \underset{\text{Carbonous oxide.}}{CO.}$$

Experiment.—For preparing carbonous oxide from oxalic acid, such an apparatus as is shown in Fig. 82, is required. The oxalic acid is placed in the flask, enough sulphuric acid is added to cover it, and the whole is heated in a cup of sand over the gas-flame. The gases as evolved pass into a washing-bottle containing a solution of potassium hydrate, in bubbling through which, the carbonic di-oxide is absorbed; the carbonous oxide thus purified, may be collected over water.

305. Properties.—Carbonous oxide is a colorless gas having a peculiar suffocating odor. It is not condensable to a liquid, and requires 40 times its volume of water for solution. Its specific gravity is 0·969. It is readily combustible, burning in the air or in oxygen with the characteristic lambent blue flame often seen playing over a freshly-fed anthracite fire. It unites directly with chlorine in sunlight, forming carbonyl chloride, or phosgene gas. It is totally irrespirable, being

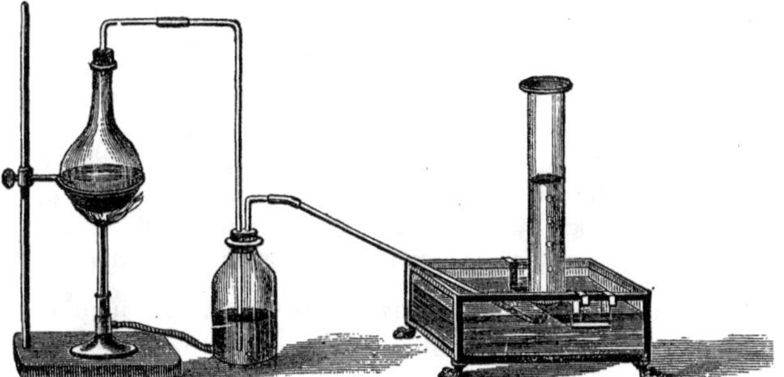

Fig. 82. Preparation of Carbonous Oxide.

an active narcotic poison, one per cent in the air proving fatal. The ready passage of this gas through heated cast-iron, and its consequent presence in apartments heated by stoves or furnaces of this metal, has been assumed as the cause of serious disease in many instances. With sufficient ventilation, however, no danger from this cause need be apprehended.

COMBUSTION.

306. Definition.—In the wide sense of the term, combustion is an energetic chemical action, attended with light and heat. It is commonly restricted, however, to the direct union of a substance with oxygen. Two substances at least, are concerned in every combustion: the combustible, or the body which burns; and the supporter of combustion, or the gas in which the combustion takes place.—Commonly we call hydrogen a combustible body, and air a supporter of combustion; but if the atmosphere were hydrogen, then the oxygen would burn in it, and would be the combustible body.

Fig. 83. Oxygen burning in Coal-gas.

EXPERIMENT.—It is commonly said that coal-gas burns in the air; but if a jar be filled with coal-gas and a combustion-spoon containing melted potassium chlorate be introduced into it, Fig. 83, the oxygen given off will burn in the coal-gas; the latter being now the supporter.

307. Combustibles.—The substances which are burned as combustibles are very numerous. Illuminating-gas has already been mentioned. Of liquids, the vegetable oils known as rape, olive, and turpentine, the animal oils called sperm and lard, and the mineral oils derived from petroleum, may be mentioned. Of solids from the vegetable kingdom, wood and bayberry wax; from the animal, tallow, and its product, stearin; and from the mineral, paraffin and the various sorts of coals, are examples. These substances, though so different in character and origin, all agree in the fact that they contain carbon and hydrogen. Some contain oxygen in addition.

308. Combustion itself.—Bodies are burned either for purposes of warmth or illumination. The temperature at which bodies take fire in the air, differs widely; phosphorus, for example, inflames at 50°; sulphur at 260°; hydrogen at 500°: while coal-gas requires a temperature of 1,000°, and nitrogen one of 5,400°. The amount of heat and light evolved are proportional to the rapidity of the combustion. In the decay of wood, there is a true burning—called by Liebig eremacausis; but the heat produced is small. Phosphorus exposed to the air becomes luminous and slowly oxidizes; but the temperature never rises very high. When heated very hot, however, bodies undergo a rapid combustion, and evolve their maximum of heat. When the matter burning is gaseous, then the phenomenon of flame appears. The character of flame may be very well studied in that of an ordinary candle, Fig. 84. The solid matter, melted by the heat, is drawn up in the wick as a liquid, and is converted in the flame into gas. In the center of the flame, then, there is a dark cone of inflammable gas. Surrounding this there is the luminous envelop, where condensed hydrocarbons are undergoing combustion. And outside still is the portion of the flame called the mantle, faintly blue in color, composed, it is said, of burning hydrogen, and cup-shaped at its lower portion.

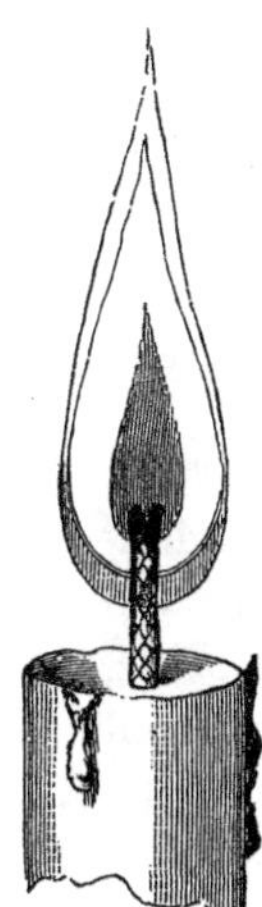

Fig. 84. Candle-flame.

EXPERIMENTS.—By means of a glass tube, the combustible gas in the center of a candle-flame may easily be led off and burned. Moreover, burning phosphorus placed in the center of a large alcohol-flame is extinguished; thus showing the absence of air there. A porcelain plate held in the luminous cone will have the condensed hydrocarbons deposited upon it; i. e., will be smoked.

In order to burn this gaseous matter in the center of the flame, Argand proposed the burner now known by his name. In this burner, air is admitted into this gaseous space by making the flame circular. Moreover, by using a chimney to produce a draft, great bril-

Fig. 86. Bunsen's burner.

liancy and steadiness are given to the flame. For heating purposes, the best burner is Bunsen's burner, Fig. 86. The gas, issuing from a small central jet, is thoroughly mixed with air, entering by the lateral openings, before it burns at the top of the tube. By this dilution, which may be made equally well with nitrogen or carbonic di-oxide, the density of the gas is so reduced that the flame is non-luminous, and no smoke is deposited on bodies held in it.

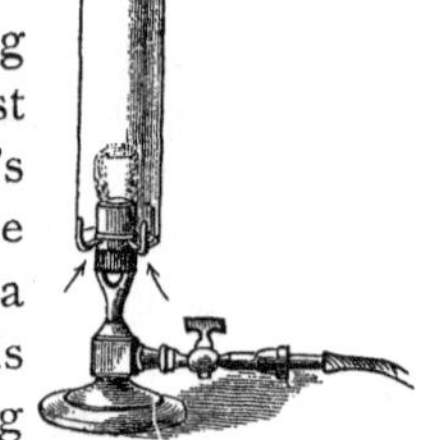
Fig. 85. Argand burner.

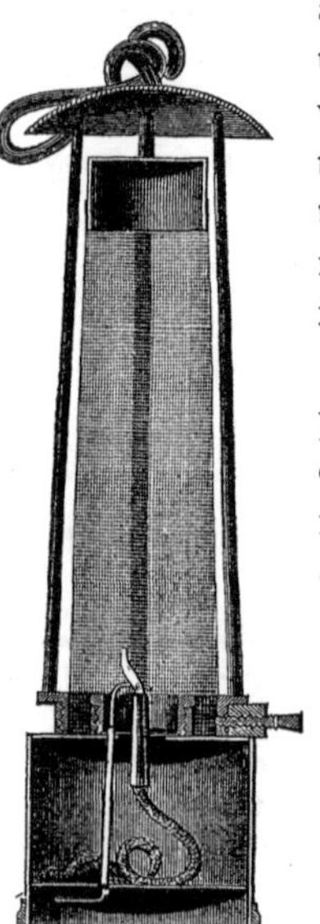
Fig. 87. Safety-lamp.

Flame, if cooled below a certain point, is extinguished; hence no flame can be propagated through a cold, fine metal tube. This was Davy's discovery, which gave rise to the safety-lamp.

Experiments. — Wire-gauze is simply a collection of small, short tubes. If a piece of such gauze be pressed upon a gas-flame, the flame will not pass through it, but will be depressed by it as if it were a solid plate. If held two inches above the jet, the gas may be lighted above the gauze, but the flame will not pass through to the jet. With two pieces of gauze the gas may be made to burn between them, but neither above the upper nor below the lower; or above and below, but not between them.

The miner's safety-lamp is represented in section in Fig. 87. It consists of a metallic lamp, the wick of which is surrounded with wire-gauze, enclosed in a frame, by which the whole may be suspended. The explosive mixture of air with the gases of the mine can enter the gauze and burn within it; but the flame cannot pass outward through the gauze, as it is thereby cooled and extinguished. Hence, with such a lamp an explosion of these gases cannot take place.

308. Products of Combustion.—The products of combustion are of two sorts: 1st, the physical products, the heat and light for which the combustion is generally produced. 2d, the chemical products, which are to be conveyed away. The heat of combustion is measured in heat-units; a heat-unit being the quantity of heat required to raise one gram of water from 0° to 1°. Thus the heat of combustion of hydrogen in oxygen is 34,462 units, and that of carbon 8,080 units; i. e., one gram of hydrogen or of carbon in burning in oxygen, would heat 34,462 or 8,080 grams of water from 0° to 1°. Knowing the quantity of carbon and hydrogen in a given fuel, therefore, it is easy to calculate its value for heating purposes.

The light given by an illuminating agent is measured by comparison with that of a standard candle, as mentioned under coal-gas. The instrument used is called a photometer.

The chemical products of combustion, since the combustibles are composed of carbon and hydrogen, are obviously carbonic di-oxide and hydrogen oxide or water.

EXPERIMENTS.—That water is produced in combustion may be shown by holding a cold dry bell-glass over a candle-flame; it will be at once bedewed with moisture. If now a little clear lime-water be shaken in the jar, it will become milky, thus proving the presence of carbonic di-oxide. The same is true of respiration; a full breath blown through a glass tube into lime-water will make it entirely white. Air which has been twice respired will also extinguish a candle.

309. Ventilation.—To carry off these effete products, an efficient ventilation is necessary. This is generally secured by a draft produced by heat, carrying all the foul air into a ventilating shaft and thence into the outer air. Air is too impure to breathe if it contains 0·10 per cent of carbonic di-oxide; but numerous other organic and inorganic impurities are produced by respiration and by combustion in our houses, which are quite as injurious to health, and which must also be removed.

EXPERIMENTS.—The principles of ventilation may be illustrated by the following experiments: If a tall bell-glass, closed at top, be placed over a stand on which are three lighted candles at different hights, Fig. 88, the heated carbonic di-oxide will accumulate in the upper part of the bell, and gradually extinguish the tapers from above downward. By removing the stopper and raising the jar just before the last flame expires, the air is renewed and the taper will be revived.

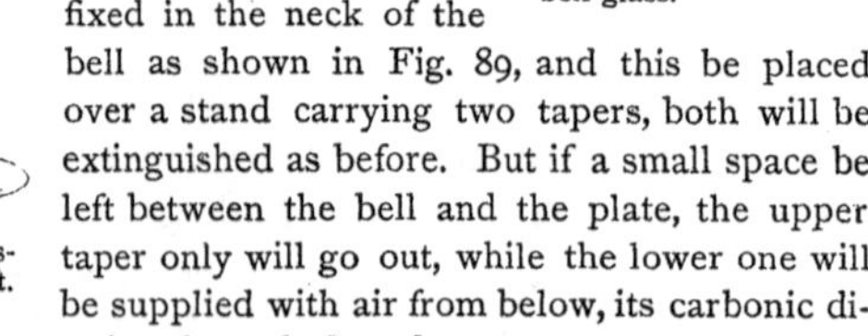

Fig. 88. Accumulation of CO_2 in upper part of bell-glass.

If a wide glass tube be fixed in the neck of the bell as shown in Fig. 89, and this be placed over a stand carrying two tapers, both will be extinguished as before. But if a small space be left between the bell and the plate, the upper taper only will go out, while the lower one will be supplied with air from below, its carbonic di-oxide and water escaping through the tube.

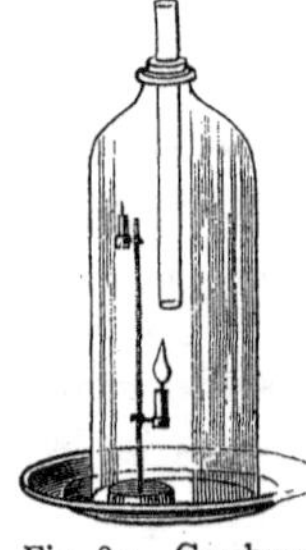

Fig. 89. Combustion with a Draft.

Again, if two large tubes, one within the other, be fixed in the neck of a bell-glass, Fig. 90, and the whole be placed over a candle-flame, the heated and effete products of combustion escape up the center tube while fresh air enters by the annular space between the two, and the candle burns actively.

Mines are usually ventilated by means of two shafts, called the upcast and the downcast shaft, respectively. Their action may be represented by Fig. 91. A square box has at each end a chimney, in one of which a candle is kept burning. The air heated by the combustion rises, while fresh air descends in the opposite chimney to supply its place. Practically, a coal fire is kept burning at the base of the upcast shaft, and by suitably arranging doors in the different parts of the mine, the whole may be thoroughly ventilated. If but one shaft is possible, then an arrangement illustrated by Fig. 92 is generally employed. A candle placed in a small bell-glass surmounted by a wide tube, will be extinguished. But if a piece of tin-plate be inserted in the tube, the foul air will pass out on one side and fresh air will enter on the other, and the candle will continue to burn. It was the taking fire of such a partition, and the consequent stoppage of ventilation, which caused the terrible disaster at the Avondale colliery.

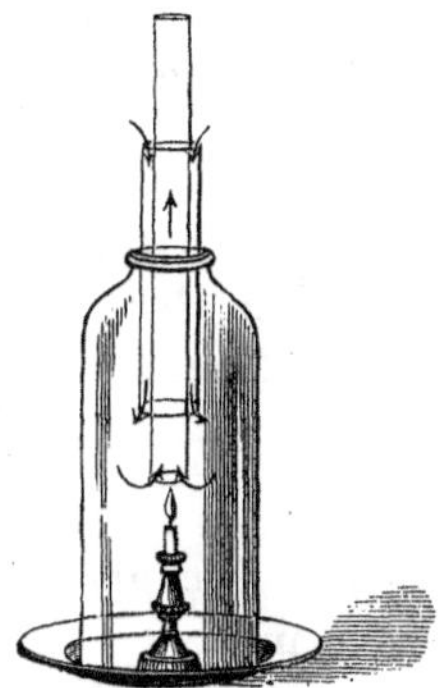

Fig. 90. Double-draft Apparatus.

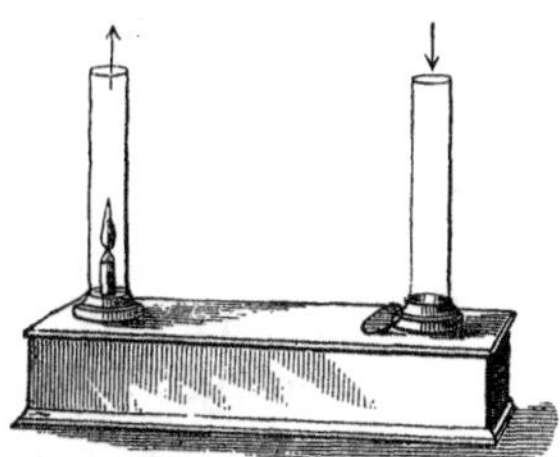

Fig. 91. Upcast and Downcast Shafts.

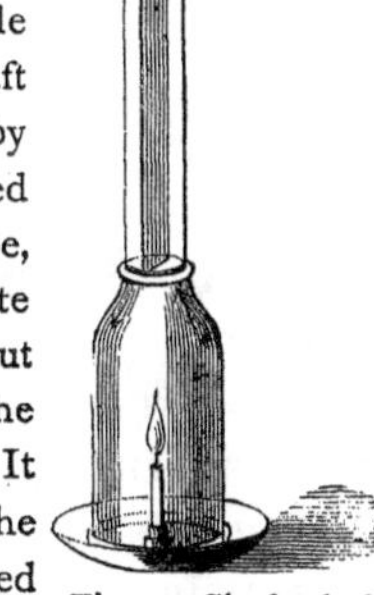

Fig. 92. Single-shaft ventilation.

CARBON AND SULPHUR.

CARBONIC DI-SULPHIDE.—*Formula* CS_2. *Molecular weight* 76. *Molecular volume* 2. *Density of vapor* 38. *One liter of carbonic di-sulphide vapor weighs* 3·40 *grams (*38 *criths).*

310. History and Preparation.—Carbonic di-sulphide was discovered by Lampadius in 1796. It is always prepared synthetically, by passing the vapor of sulphur over charcoal heated to redness :—

$$C_2 + (S_2)_2 = (CS_2)_2.$$

By agitating with lead hydrate or mercuric chloride, and re-distilling from milk of lime, it is obtained pure.

311. Properties.—Carbonic di-sulphide is a colorless, strongly refracting liquid, having, when perfectly pure, an agreeable etherial odor, like chloroform. Its specific gravity is 1·27. It boils at 46°, yielding a dense vapor. It is very volatile, evaporating rapidly in the air, with the production of great cold. It is not soluble in water. Its vapor is readily inflammable, taking fire in the air even at 150°, and burning with a blue flame, producing carbonic and sulphurous oxides. The brilliancy of its combustion with nitrogen di-oxide has been already mentioned.

Carbonic di-sulphide unites directly with alkaline sulphides and sulphydrates, producing sulpho-carbonates M_2CS_3, analogous to carbonates M_2CO_3. By decomposing ammonium sulpho-carbonate with hydrochloric acid, sulpho-carbonic acid H_2CS_3 is obtained as a reddish-brown oily liquid :—

$$(NH_4)_2CS_3 + (HCl)_2 = (NH_4Cl)_2 + H_2CS_3.$$

312. Uses.—Carbonic di-sulphide is used in the arts to dissolve phosphorus, iodine and sulphur; the latter particularly in vulcanizing india-rubber. It is largely used also as a solvent for resins and bitumens; and of late years has been prepared on an immense scale for the extraction of the various fatty and essential oils.

CARBON AND NITROGEN.

CYANOGEN.—*Formula* C_2N_2. *Symbol* Cy. *Molecular weight* 52. *Molecular volume* 2. *Density* 26. *One liter weighs* 2·32 *grams (*26 *criths).*

313. History.—Cyanogen was discovered by Gay-Lussac in 1815. It was the first compound radical isolated, and its discovery marks an era in the science of chemistry. Its name is from κυάνεος, dark-blue, it being a constituent of the well-known pigment, prussian-blue.

314. Preparation.—Cyanogen is usually prepared by heating the cyanide of gold, silver, or mercury:—

$$Hg''Cy_2 = Hg + Cy_2.$$

EXPERIMENT.—Instead of using the somewhat rare mercuric cyanide, a mixture of two parts of thoroughly dried potassium ferrocyanide and three parts mercuric chloride may be advantageously substituted. The mixture is placed in a flask of hard glass, Fig. 93, upon a sand-bath, and heated intensely by means of the double-draft gas-burner. The gas as it is evolved, may be collected by displacement, the pungent odor indicating when the jar is full.

315. Properties.—Cyanogen is a colorless gas, with a penetrating, pungent odor, resembling that of peach-blossoms. Its specific gravity is 1·806. Cooled to —20·7°, or subjected to a pressure of four and a half atmospheres at 15°, it condenses to a colorless, highly refractive liquid, of specific gravity 0·866, which freezes at $-34°$ to a transparent, ice-like solid. Cyanogen gas is soluble in one-fourth of its volume of water, and in one-twentieth of its volume of alcohol. It takes fire readily in the air, burning with a characteristic purple-red flame, producing carbonic di-oxide and nitrogen.

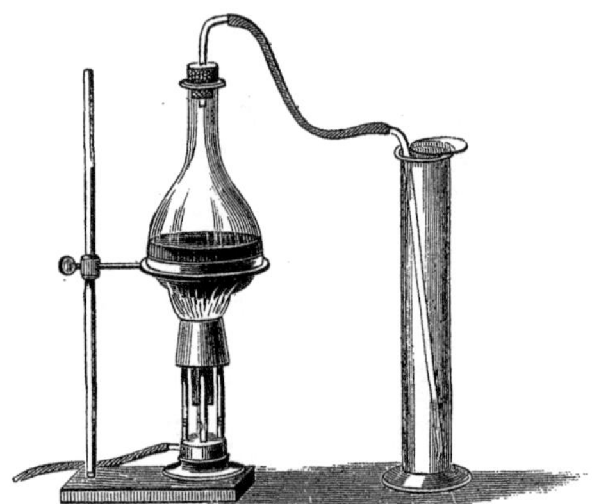

Fig. 93. Preparation of Cyanogen.

Free or molecular cyanogen is composed of two atoms of the cyanogen radical, Cy_2 being analogous to Cl_2. Moreover, atomic cyanogen acts precisely like an elemental monad, forming compounds corresponding to the chlorides, thus :—

Potassium chloride	KCl.	Potassium cyanide	KCy.
Hydrochloric acid	HCl.	Hydrocyanic acid	HCy.
Hypochlorous acid	$HClO$.	Cyanic acid	$HCyO$.

§ 2. SILICON.

Symbol Si. *Atomic weight* 28. *Equivalence* IV. *Density* 28 *(?)*. *Molecular weight* 56 *(?)*. *Molecular volume* 2. *One liter of silicon-vapor weighs* 2·5 *grams (*28 *criths) (?)*.

316. History and Occurrence.—Silicon was first obtained pure by Berzelius in 1823. It does not occur free in nature, but exists abundantly in combination with oxygen, forming the well-known substance, quartz. Combined with oxygen, and also with potassium, aluminum, and other metals, it constitutes a large portion of the rocks which make up the solid crust of the earth.

317. Preparation.—Silicon may be prepared by the action of sodium upon potassium fluo-silicate :—

$$K_2SiF_6 + Na_4 = (KF)_2 + (NaF)_4 + Si.$$

318. Properties.—Silicon is thus obtained as an amorphous, nut-brown, lusterless powder, which may be crystallized by solution in melted zinc, in long needles made up of adhering regular octahedrons ; or in melted aluminum, in flat octahedral plates. In this form it is a dark iron-gray solid, with a metallic luster and a specific gravity of 2·49. It is hard enough to scratch glass, and melts at a temperature above the melting point of iron. Heated in the air, it takes fire and burns, producing silicic oxide.

SILICON AND HYDROGEN.

HYDROGEN SILICIDE. — *Formula* H_4Si. *Molecular weight* 32.

319. History and Preparation.—Hydrogen silicide was

first observed by Wöhler and Buff in 1857; but it was first obtained pure in 1867, by Friedel and Ladenburg.

It may be obtained, mixed with hydrogen, by the electrolysis of a solution of sodium chloride, a plate of aluminum containing silicon, being made the positive electrode. A better process is to decompose magnesium silicide by hydrochloric acid :—

$$Mg_2Si + (HCl)_4 = (MgCl_2)_2 + H_4Si.$$

It is obtained perfectly pure by the action of sodium upon tri-ethyl silico-formate.

EXPERIMENT.—For preparing magnesium silicide, 40 parts fused magnesium chloride, 35 parts dried sodium fluo-silicate, 10 parts fused sodium chloride and 20 parts of sodium cut into small pieces, are well mixed together, and thrown into a red-hot hessian crucible, which is then covered and heated until the sodium ceases to burn. When cold, a dark layer of the impure silicide will be found at the bottom. It is detached and preserved in a tight bottle.

Fig. 94. Preparation of Hydrogen Silicide.

To obtain hydrogen silicide, the coarsely pulverized slag thus made, is placed in a wide-mouthed bottle, Fig. 94, through the cork of which passes a funnel tube for adding the acid, and a wide delivery tube for the escape of the gas. The bottle is then filled with cold water —recently boiled to expel the air—and placed by the water-cistern as shown in the figure. Upon pouring concentrated hydrochloric acid down the funnel tube—taking especial care that it carries in no air-bubbles—hydrogen silicide gas is evolved and may be collected for use.

320. Properties.—Hydrogen silicide is a colorless gas, which, when mixed with hydrogen, is spontaneously inflammable in the air, yielding white clouds of silicic oxide. Burned from a jet, it gives a brilliant white flame, which deposits a brown layer of silicon on a piece of porcelain held in it. When the tube conveying the gas is heated, a mirror-like deposit of silicon takes place within it. Hydrogen silicide is not soluble in water. Passed into cupric sulphate, or silver nitrate, it throws down cupric and silver silicides. It is decomposed by potassium hydrate, one volume of the gas yielding four volumes of hydrogen :—

$$SiH_4 + (HKO)_4 = K_4SiO_4 + H_8.$$

One half of this hydrogen comes from the silicide ; each molecule must therefore contain two molecules of hydrogen, or four atoms.

SILICON AND OXYGEN.

SILICIC OXIDE.—*Formula* SiO_2. *Molecular weight* 60.

321. Occurrence.—Silicic oxide or silica, occurs abundantly in nature in the pure and crystallized form known as quartz or rock-crystal ; and more or less deeply colored in the minerals called amethyst, jasper, agate, chalcedony, and carnelian. A beautiful, and probably polymeric form, is known as opal. As sandstone, it forms a rock-material in geology. Its combinations with metallic oxides, called silicates, form by far the most numerous class of mineral substances. Silicic oxide also occurs dissolved in many natural waters, particularly those of thermal springs ; it stiffens the stems of the cereal grains, and has been also found in animal tissues.

322. Preparation. — Silicic oxide may be prepared

either by the oxidation of silicon, as when it burns in the air, or by the dehydration of silicic acid:—

$$H_2SiO_3 - H_2O = SiO_2.$$

323. Properties.—Silicic oxide, as usually obtained, is a white amorphous powder, though in nature it occurs often in the form of hexagonal prisms, crowned by six-sided pyramids, Fig. 95. It has a specific gravity of 2·6, is so hard as readily to scratch glass, and is fusible only by the oxyhydrogen flame. It is but slightly soluble in water and is unattacked by acids, except the hydrofluoric. Fused with salts of the alkali-metals, it combines with the basic oxide, forming a silicate, and sets the negative or acid oxide free.

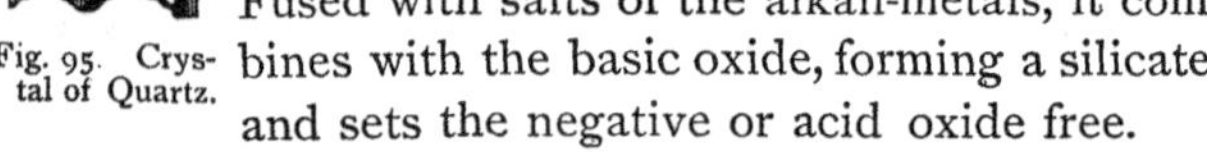

Fig. 95. Crystal of Quartz.

324. Silicic Acid. — *Ortho* $H_4Si^{IV}O_4$. *Meta* $H_2Si^{IV}O_3$. Ortho-silicic acid may be prepared by the action of water on silicon fluoride, or by decomposing a solution of an alkaline silicate by an acid. It is obtained as a gelatinous precipitate, losing water when dried, and producing meta-silicic acid. By dialysing a solution of an alkaline silicate in hydrochloric acid, the acid passes through the membrane, and there is left in the dialyser an aqueous solution of silicic acid, which may be concentrated by boiling in a flask until it contains 14 per cent. It is a tasteless, limpid liquid, is slightly acid, and becomes a jelly on standing. On evaporation, it leaves meta-silicic acid. A third silicic hydrate, called para-silicic or di-silicic acid, $H_6Si_2O_7$, is known.

The silicates are classified by Dana as uni-silicates, M'_4SiO_4, corresponding to ortho-silicates, and bi-silicates, M'_2SiO_3, corresponding to meta-silicates. Some of the intermediate varieties, such as wernerite [Na, Ca, Al]$''_3$ Si_2O_7, are di- or para-silicates; others, as lepidolite [K, Li, Mn, Al]$''_4$ Si_3O_{10}, and labradorite, [Ca, Na, Al]$''_4$ Si_3O_{10}, are derived from tri-silicic acid, $H_8Si_3O_{10}$. Beside these, there is a third class of silicates—Dana's sub-silicates—which are meta-aluminic silicates, derived from meta-aluminic base $H_4Al_2{}^{VI}O_5$—by replacing the H_4 by Si^{IV}. In this way, the mineral cyanite, $SiAl_2O_5$, is produced. Potassium and sodium meta-silicates are soluble in water, and their solutions are used in the arts under the name of soluble glass. Potassio-lead silicate is known in commerce as flint-glass, and sodio-calcium silicate as crown or window glass.

§ 3. Tin.

Symbol Sn. *Atomic weight* 118. *Equivalence* II *and* IV. *Molecular weight* 236 *(?)*.

325. History and Occurrence.—Tin has been known from the remotest antiquity. It is spoken of by Moses (Numbers, xxxi, 22); and Homer mentions it in the Iliad under the name *κασσίτερος*. Much of the brass of the ancients was a copper and tin bronze, the tin being obtained from Cornwall; whence Herodotus speaks of the British Isles as the *κασσιτερίδες* or tin-islands. The principal ore of tin is stannic oxide, known as the mineral cassiterite. It occurs in veins running through ancient rocks—vein or mine-tin—and also in the beds of water-courses, from the disintegration of these rocks—stream tin. The principal localities of tin-ore are Cornwall, England, and Banca and Malacca, India. It has been found in New Hampshire and in California in this country.

326. Preparation and Properties.—The ore is pulverized, roasted, and washed, and is then smelted with charcoal, which removes the oxygen. It is refined by melting and thrusting into it a stick of wet wood; the impurities rise to the surface and are skimmed off. The tin is then ladled into moulds, the upper layers being the purest. It comes into commerce as grain-tin, in irregular fragments produced by allowing the hot ingots to fall from a height; and as block-tin, a less pure form, in ingots. The purest tin is imported from the island of Banca, and is known as straits-tin.

Tin is a soft, brilliant-white metal, of specific gravity 7·29. It is dimorphous, crystallizing in forms belonging to the isometric and the quadratic systems. It is very malleable, and may be beaten into leaves one fortieth of a millimeter thick; at 100° it is ductile, and may be drawn into wire. Its tenacity, however. is small. It crackles when a bar of it is bent, producing what is known as the cry of tin. It has a peculiar odor, and is a good conductor of heat and electricity. It melts at 230°, and is not volatile. Heated in the air, it burns readily to oxide, though it retains its luster in air at ordinary temperatures. Acids attack it readily.

327. Uses.—Tin is used in the arts for making tin foil, for covering iron in the preparation of tin-plate, and for alloys. Pewter, brittania, queen's metal, and solder, are alloys of tin and lead, containing sometimes a little antimony and bismuth. Bell-metal, gun-metal, bronze, and speculum-metal are essentially alloys of tin and copper.

TIN AND CHLORINE.

STANNIC CHLORIDE. — *Formula* $SnCl_4$. *Molecular weight* 260.

328. Preparation and Properties.—Stannic chloride,—known to the alchemists as Liquor fumans Libavii,—may be obtained by the direct action of chlorine gas upon tin, or by distilling mercuric chloride with tin-filings :—

$$(HgCl_2)_2 + Sn = Hg_2 + SnCl_4.$$

It is a colorless, fuming liquid, of specific gravity 2·28, which boils at 115° ; its vapor density is 131. It unites with water readily, evolving heat, and forming two crystalline hydrates. With alkali-chlorides it forms definite compounds, the potassium salt being K_2SnCl_6. It is used in dyeing.

STANNOUS CHLORIDE.—*Formula* $SnCl_2$. *Molecular weight* 189.

329. Preparation and Properties. — Stannous chloride may be prepared by distilling tin-filings with mercurous chloride, or by the action of heat upon its hydrate. It is a grayish-white translucent solid, which melts at 250°, and may be distilled. By solution in water and evaporation, large colorless monoclinic prisms are produced, having the composition $SnCl_2$, 2aq. These crystals are known as the "tin-salt" of the dyer, for whose use they are commonly made by dissolving metallic tin in hydrochloric acid. Stannous chloride is used in the laboratory as a reducing agent.

TIN AND OXYGEN.

STANNIC OXIDE.—*Formula* $Sn^{IV}O_2$. *Molecular weight* 150.

330. Preparation and Properties. — Stannic oxide occurs native as the mineral cassiterite, or tin-stone, crystallized in square prisms, terminated by the faces of the square octahedron. It may be prepared by burning the metal in air, or by igniting either of the hydrates. It is then obtained as a white powder, of specific gravity 6·6, insoluble in all acids but hydrofluoric. Owing to its hardness it is used for polishing glass under the name of putty-powder. When fused with alkali-hydrates, it forms stannates.

331. Stannic Acids.—*Ortho* H_4SnO_4 *and Meta* H_2SnO_3. Ortho-stannic acid is precipitated from alkaline stannates by acids, or from stannic chloride by ammonia, as a gelatinous mass, which loses water on drying in vacuo, and becomes meta-stannic acid. When tin is oxidized by nitric acid, a polymeric form of meta-stannic acid is produced, which when dried at 100° has the formula $H_{10}Sn_5O_{15}$. Sodium meta-stannate Na_2SnO_3, 3aq., made by fusing native tin-stone with sodium hydrate, is used as a mordant in dyeing.

STANNOUS OXIDE.—*Formula* $Sn''O$. *Molecular weight* 134.

332. Preparation and Properties. — Stannous oxide is obtained when stannous oxalate is heated in close vessels. It is a black powder of specific gravity 6·6, crystallizing in the isometric system, and combustible when heated in the air. With water it forms a hydrate which absorbs oxygen gradually from the air, and passes into stannic acid.

TIN AND SULPHUR.

STANNIC SULPHIDE. — *Formula* $Sn^{IV}S_2$. *Molecular weight* 182.

333. Preparation and Properties.—Stannic sulphide is prepared by heating together tin-amalgam, sulphur, and ammonium chloride (sal-ammoniac). A golden-yellow crystalline powder, having a metallic luster, and a specific gravity of 4·6, is left in the vessel. This substance was known to the alchemists under the name of aurum musivum, or mosaic gold. It is used as a bronze-powder.

RELATIONS OF THE CARBON GROUP.

334. Carbon, Silicon, Titanium, and Tin, are closely related to each other in the gradation of their physical properties as well as in the similarity of their chemical compounds. Their tetrad character is well marked, since they all form tetra-chlorides, di-oxides, and di-sulphides; and since carbon and silicon form also tetra-hydrides. Moreover, carbon and tin form un-saturated, or rather self-saturated compounds, in which they act as dyads; carbon thus forming carbonous oxide, and tin, stannous chloride, oxide, and sulphide. The elements of this group are also capable of entering into combination as double atoms, with an equivalence of six; forming the chloride of carbon C_2Cl_6, the oxide of titanium Ti_2O_3, and the sulphide of tin Sn_2S_3, for example. Finally, they all form normal ortho-acids which are tetra-basic, and derived meta-acids, which are dibasic.

EXERCISES.

§ 1.

1. Mention the forms in which carbon occurs in nature.
2. How are they proved to be chemically identical?
3. One kilogram of carbon in burning will evaporate how much water?
4. How is methane produced naturally? How made artificially?
5. What volume of methane from a kilogram of sodium acetate?
6. One gram of alcohol yields what volume of ethylene?
7. A cubic meter of ethylene contains what volume of H?
8. For what reason is ethylene written C_2H_4 and not CH_2?
9. What volume of O is required to burn 250 grams of ethylene?
10. How is acetylene prepared? What are its properties?
11. Describe the process for making coal-gas.
12. One kilogram of C in burning gives what weight of CO_2?
13. What weight of $KClO_3$ will completely burn 5 grams of C?
14. What volume of air is required to burn one liter of marsh-gas, of ethylene, and of acetylene? What volume of CO_2 is produced in each case?
15. What volume of carbonic di-oxide will be produced by burning a kilogram of cannel coal, 85·81 per cent of which is carbon?
16. One c. c. of marble (sp. gr. 2·7) contains what vol. of CO_2?
17. How many kilograms of carbon in 1000 kilograms of chalk?
18. What volume of CO_2 must be passed over charcoal to yield a liter of CO? What is the increase in weight?
19. 25 grams oxalic acid yield what volume of CO at 10° and 740 m. m. pressure?
20. To burn one gram of CS_2 requires what volume of O?
21. What volume of air is required to burn one liter of CN?
22. What are the analogies of cyanogen?

§ 2.

23. Give the method of preparation and the properties of silicon.
24. Into what classes are the silicates divided? Illustrate.

§ 3.

25. How does tin occur in nature? How is it obtained?
26. What weight of "tin salts" will 250 kilograms of tin yield?

CHAPTER SEVENTH.

THE IRON GROUP.

§ 1. CHROMIUM.

Symbol Cr. *Atomic weight* 52·5. *Equivalence* II, IV, *and* VI. *Also a pseudo-triad* $(Cr_2)^{VI}$.

335. History and Occurrence.—Chromium was first recognized as a distinct substance by Vauquelin in 1797, in a native lead chromate from Siberia. Its name comes from χρῶμα, color, because most of its compounds are brilliantly colored. It occurs in nature in the mineral chromite, or chrome-iron, a ferroso-chromic oxide. Also as lead chromate in the mineral crocoite. It forms the coloring matter of the emerald, and has been found in meteoric irons.

336. Preparation and Properties. — Chromium is prepared by reducing its oxide by charcoal; or better, by reducing the chloride by zinc. By the former method, it is obtained as a steel-gray mass, infusible, and extremely hard; by the latter, as a gray-green, glistening powder, composed of minute tetragonal octahedrons, of specific gravity 6·8. It is unaltered when heated in dry air, but burns in oxygen. It is not magnetic.

CHROMIUM AND CHLORINE.

337. Chromic Chloride. — *Formula* Cr_2Cl_6. Chromic chloride is obtained by passing chlorine gas over ignited pellets of chromic oxide and lamp-black. A sublimate of micaceous scales, beautiful peach-blossom in color, and of an unctuous feel, is thus produced, which is insoluble in water unless a trace of chromous chloride is present; then it dissolves readily. It is unaltered by ordinary re-agents. On prolonged boiling with water it dissolves, forming a green solution, containing a hydrate. A soluble violet modification is produced by heating the green hydrate in a current of hydrochloric acid gas. Its solution becomes green on boiling. The chlorine of the violet solution is completely precipitated by silver nitrate; that of the green solution but partially.

338. Chromous Chloride.—*Formula* $CrCl_2$. Chromous chloride is prepared by reducing chromic chloride by hydrogen, at a gentle heat. It is a white, crystalline substance, soluble in water, forming a blue solution, which by absorption of oxygen rapidly becomes green.

339. Chromic Per-fluoride.—*Formula* CrF_6. Chromic per-fluoride is obtained by distilling lead chromate with calcium fluoride (fluor spar) and sulphuric acid:—

$$PbCrO_4 + (CaF_2)_3 + (H_2SO_4)_4 = PbSO_4 + (CaSO_4)_3 + (H_2O)_4 + CrF_6.$$

An orange vapor is evolved, which condenses to a blood-red liquid, boiling at a temperature but little above that of the air, fuming in contact with moist air and decomposed by water, forming hydrofluoric and chromic acids.

CHROMIUM AND OXYGEN.

340. Chromic Tri-oxide. — *Formula* $Cr^{VI}O_3$. Chromic tri-oxide is obtained by mixing a saturated solution of potassium di-chromate with its own volume of strong sulphuric acid. On cooling, splendid crimson needles of the trioxide crystallize out, which may be dried on a porous tile. It is deliquescent in damp air, and is decomposed by a heat of 250°. It is a powerful oxidizing agent; alcohol poured upon it is at once inflamed, and ammonia gas reduces it with incandescence.

341. Chromic Acid.—*Formula* H_2CrO_4. By solution of chromium trioxide in water, an acid liquid is obtained which contains chromic acid, but which is decomposed on evaporation, yielding only chromic trioxide again. The salts of chromic acid, called chromates, are numerous and important. The ortho-chromates of bismuth Bi'''_2CrO_6, and of mercury Hg''_3CrO_6, the mono-meta-chromate of lead, Pb''_2CrO_5, the di-meta-chromates of sodium Na_2CrO_4, and of barium $Ba''CrO_4$, and the di-chromate of potassium $K_2Cr_2O_7$, are all well-known compounds. Potassium di-chromate is used in dyeing and in calico-printing.

342. Perchromic Acid.—*Formula* $H_2Cr_2O_8$ (?). By the action of hydrogen peroxide upon an acid solution of potassium chromate, a bright-blue liquid is obtained which evolves oxygen with effervescence, and becomes green. By agitation with ether, the blue substance is dissolved; and on standing, it forms a bright-blue etherial layer on the surface of the liquid. This reaction is a delicate test for hydrogen peroxide or for a chromate.

343. Chromic Oxide.—*Formula* $Cr_2^{VI}O_3$. Chromic oxide may be produced by igniting its hydrate, or by decomposing the trioxide or a dichromate by combustibles. By passing chromyl chloride vapor through a red-hot tube, rhombohedral crystals of chromic oxide are obtained, greenish-black in color, having a specific gravity of 5·21, and hard enough to scratch glass. It is generally produced in the form of an amorphous bright-green powder, which after ignition is insoluble in acids. It is used to color bank-notes green.

Chromic oxide may act as a positive or a negative oxide according to the oxide with which it unites. With the strongly negative sulphuric oxide, for example, it forms chromium sulphate, $Cr_2(SO_4)_3$; while with calcium or magnesium oxide, compounds called calcium or magnesium chromites, $CaCr_2O_4$ or $MgCr_2O_4$, are obtained. The best known of these is $FeCr_2O_4$, ferrous chromite, or native chromic iron. Chromium sulphate exists in solution in two different modifications, one green, the other violet; with potassium sulphate, the latter yields a double sulphate, which, crystallizing in violet-red regular octahedrons with twenty-four molecules of water, is known as potassio-chromium alum, $K_2Cr_2(SO_4)_4$, 24aq. Ortho-chromic hydrate, $H_6Cr_2O_6$. 4aq., is precipitated by ammonium hydrate from boiling solutions of chromic salts; and mono-meta-chromic hydrate, $H_4Cr_2O_5$, is used as a pigment under the name of Pannetier's green.

344. Chromous Oxide.—*Formula* $Cr''O$. Chromous oxide is known only in the form of hydrate, produced by precipitating chromous chloride by potassium hydrate. It acts as a basic oxide, yielding chromous salts.

345. Chromyl Chloride.—*Formula* $(CrO_2)''Cl_2$. Chromyl chloride is prepared by distilling a mixture of sodium chloride and potassium dichromate with sulphuric acid. It is a blood-red liquid, having a specific gravity of 1·71, and boiling at 118°. Its relation to the chromates is as follows:—

$CrO_2\left\{\begin{matrix}OH\\OH\end{matrix}\right.$ $CrO_2\left\{\begin{matrix}OH\\OK\end{matrix}\right.$ $CrO_2\left\{\begin{matrix}OK\\OK\end{matrix}\right.$ $CrO_2\left\{\begin{matrix}Cl\\OK\end{matrix}\right.$ $CrO_2\left\{\begin{matrix}Cl\\Cl\end{matrix}\right.$

Hydrogen chromate. *Hydro-potassium chromate* *Potassium chromate.* *Potassium chloro-chromate.* *Chromyl chloride.*

Potassium chloro-chromate crystallizes from a hot solution of potassium di-chromate in hydrochloric acid, on cooling.

§ 2. MANGANESE.

Symbol Mn. *Atomic weight* 55. *Equivalence* II, IV, *and* VI. *Also a pseudo-triad*, $(Mn_2)^{VI}$.

346. History and Occurrence.—Manganese was discovered by Scheele and Bergmann in 1774, in a mineral known as braunstein. Owing to its being confounded with magnetic iron, this mineral had received the Latin name of this substance, magnesia nigra ; whence the name magnesium first given to the new metal obtained from it. This name was afterward changed to manganesium, to distinguish it from the true magnesium, obtained from the magnesia alba. Manganese occurs somewhat abundantly in nature, combined principally with oxygen. The mineral pyrolusite is manganese dioxide ; hausmannite is manganoso-manganic oxide ; and manganite is manganic hydrate. Manganese sulphide, arsenide, carbonate and silicate are also known as minerals.

347. Preparation and Properties. — Manganese is obtained by reducing its oxide by charcoal at a high temperature. It is a grayish-white, hard metal, resembling cast-iron, and very brittle. It is feebly magnetic and has a specific gravity of 8. It oxidizes readily in the air, and dissolves easily in acids. It forms a remarkably beautiful alloy with copper.

MANGANESE AND CHLORINE.

348. Manganic Chloride.—*Formula* Mn_2Cl_6. Manganic chloride is a very unstable compound, obtained by dissolving manganic oxide in hydrochloric acid at a low temperature. It is a brown liquid, readily evolving chlorine and becoming manganous chloride.

349. Manganous Chloride.—*Formula* $MnCl_2$. Manganous chloride, obtained by heating the hydrated compound, or by the direct action of chlorine on manganese, is a pale rose-colored deliquescent mass, which dissolves readily in water, forming a definite hydrate. The same hydrate is formed whenever manganese oxide or carbonate is dissolved in hot hydrochloric acid.

MANGANESE AND OXYGEN.

350. Manganic Tri-oxide.—*Formula* $Mn^{VI}O_3$. Neither manganic tri-oxide nor its corresponding hydrate, manganic acid, is known in the free state. Their salts, however, called manganates, are well-known compounds. Potassium manganate is produced whenever manganese compounds are heated with potassium hydrate or carbonate. A deep green mass results, which, dissolved in water and evaporated in vacuo, affords dark-green crystals, isomorphous with potassium sulphate. The manganates are all unstable, passing readily into permanganates and depositing manganese di-oxide.

351. Permanganic Acid.—*Formula* $H_2Mn_2O_8$. By distilling at 60° or 70°, a mixture of potassium permanganate and sulphuric acid, violet vapors are evolved which condense into a greenish-black liquid, asserted to be pure permanganic acid. It is deliquescent, dissolves in water with a violet color, and detonates on heating. Its salts, the permanganates, are more stable than the manganates. Potassium permanganate is prepared by removing a portion of the base from the manganate, by passing through its solution a current of chlorine gas:—

$$(K_2MnO_4)_2 + Cl_2 = (KCl)_2 + K_2Mn_2O_8.$$

The color of the liquid changes from green to purple-red, and on evaporation yields dark purple-red orthorhombic crystals, soluble in sixteen times their weight of water. Permanganates act as strongly oxidizing agents, and are used extensively as disinfectants.

352. Manganic Oxide.—*Formula* $Mn_2^{VI}O_3$. Manganic oxide is produced when manganic hydrate or manganese di-oxide is heated to low redness. It is a black powder, uniting with strong acids to form salts. Manganic sulphate forms an alum with potassium sulphate, having the formula $K_2Mn_2(SO_4)_4$, 24aq. With basic oxides, manganic oxide forms manganites, analogous to the chromites. Manganic hydrate, $H_2Mn_2O_4$, exists native as the mineral manganite.

353. Manganese Di-oxide.—*Formula* $Mn^{IV}O_2$. The dioxide of manganese occurs native as the mineral pyrolusite, in steel-gray orthorhombic prisms, of specific gravity 4·9. It is produced whenever a lower oxide is heated with free access of air. When heated, it gives off oxygen, and is extensively used in the arts as an oxidizing agent at high temperatures. Heated with sulphuric acid it evolves oxygen and forms manganous sulphate; with hydrochloric acid, it evolves chlorine.

354. Manganous Oxide. — *Formula* $Mn''O$. Manganous oxide is obtained by igniting manganous carbonate or oxalate in an atmosphere of hydrogen. It is a grayish-green powder, which has been obtained crystallized in emerald-green regular octahedrons. Its hydrate is thrown down as a white precipitate on adding a solution of potassium hydrate to one of a manganous salt. It quickly becomes brown on exposure to the air. Manganous oxide unites directly with negative oxides to form the manganous salts. Manganous sulphate, $MnSO_4$, manganous carbonate, $MnCO_3$, and manganous silicate, $MnSiO_3$, are examples. They all have a delicate pink color.

§ 3. Iron.

Symbol Fe. *Atomic weight* 56. *Equivalence* II, IV *and* VI. *Also a pseudo-triad,* $(Fe_2)^{VI}$.

355. History and Occurrence.—Iron is one of the most important, as it is one of the most abundant, of metals. It has been known from the earliest historic times, Tubal Cain being an artificer in this metal. Even in pre-historic times, implements made of it seem to have been used. It is doubtful whether it occurs native; the native iron found on the earth's surface containing generally nickel, and being of meteoric origin. Its ores however, are very numerous. Among these may be mentioned: ferroso-ferric oxide or magnetite, Fe_3O_4; ferric oxide or hematite, Fe_2O_3; ferric hydrate, or limonite, $H_6Fe_4O_9$; and ferrous carbonate or siderite, $FeCO_3$. It occurs also in numerous other minerals, and in vegetables and animals.

356. Preparation and Properties.—On the large scale in the arts, iron is produced either from the native oxide, or from the artificial oxide obtained by roasting the native carbonate or hydrate. Alternate layers of the ore, of the fuel, and of limestone, are placed in an enormous furnace, shaped like a double cone interiorly, and forty to sixty feet in hight; whence it is commonly called the "high furnace." A powerful blast of hot air enters at the bottom, and the combustible matter, at the high temperature produced, removes the oxygen from the ore, thus reducing it to the metallic state, according to the equation:—

$$Fe_2O_3 + C_3 = Fe_2 + (CO)_3.$$

The limestone unites with the silica and other impurities present, forming an easily fusible silicate, which collects above the melted iron and is drawn off as slag. After the iron is reduced to the metallic state, it takes up more carbon, becomes fusible, melts, and runs down to the bottom of the furnace, accumulating in a narrow cylinder called the crucible. When this is full, it is tapped by driving in the clay plug which closes the opening, and the melted iron runs down a suitable channel into moulds made in the sand for its reception. The manufacture of cast-iron in the high- or blast-furnace is a continuous operation; the materials are constantly added above, the slag and melted iron are drawn off from below, usually twice a day; and this is kept up until the furnace wears out. The iron thus made is known as cast- or pig-iron. Three varieties are distinguished; white pig-iron, which is hard, brittle and crystalline, uniformly brilliant and white, of specific gravity 7·5, and readily fusible; gray pig-iron, which is granular in structure, gray in color, very soft, of specific gravity 7·1, and difficult of fusion; and mottled pig-iron, which has intermediate properties, but is stronger than either. The gray iron has generally least carbon, its color being due to a separation of a part of this carbon as graphite, during the cooling. Hence the same metal suddenly cooled, as when "chilled" or cast in iron moulds, may be hard and white; and when cooled slowly in sand, be soft and gray. White iron, especially the variety known as spiegel-eisen, which contains manganese, contains nearly 6 per cent of carbon; gray iron contains from 2 to 5 per cent.

Iron is refined, or converted into wrought-iron, by burning out the carbon and silicon, as well as the impurities sulphur and phosphorus. This is effected usually by the process called "puddling," or "boiling." The pig-iron is piled up on the floor of a reverberatory furnace, in contact with some of the purer ores. On lighting the fire, it melts, and is continually stirred, to mix it thoroughly with the oxide. Gradually, the carbon and silicon are oxidized, the former escaping as gaseous carbonous oxide, the latter being retained as silicate of iron in the slag; until finally the iron becomes pasty and adheres together in spongy masses. These are collected into balls of 25 to 30 kilograms weight, and compacted, first by working between powerful jaws called squeezers, and then between rollers, by which the slag is pressed out, and the iron is made into "muck bar." The puddled bar is cut into short pieces, made into bundles, heated, and again passed through the rolls; the operation being repeated till the wrought iron is sufficiently pure. By this process the carbon is reduced to one-half of one per cent, sometimes to even less, and the other foreign matters to mere traces. If the iron retain phosphorus, it is brittle when cold, and is called "cold-short;" if it contain sulphur, it is brittle when hot, or "red-short." The iron thus obtained is bluish-gray in color, is fibrous in structure, and has a specific gravity of 7·3 to 7·9.

Pure iron may be prepared from the best commercial varieties, piano-forte wire, for example, by fusing them with pure iron oxide, beneath a layer of glass, to keep out the air, in a clay crucible. It is brilliant silver-white in color, softer than wrought-iron, capable of receiving a high polish, strongly magnetic, of specific gravity 7·8, and crystallizes in the regular system. When obtained by electrolysis, its specific gravity is 8·1. Iron is also prepared for pharmaceutical uses, by reducing its oxide by hydrogen, at a red heat. It is then obtained as a black powder, burning when heated in the air. In its purest commercial form, iron has a tenacity superior to that of any other metal, except nickel and cobalt; its ductility is also very great, and when heated, it may be rolled into sheets scarcely thicker than paper. At a full red heat it becomes pasty like wax, and may then be welded. It melts above 2,000°.

Steel is iron which contains from 0·6 to nearly 2·0 per cent of carbon. It is therefore intermediate in this respect between cast- and wrought-iron. Two methods are in use for making steel. In the one, the wrought-iron bar is heated with charcoal, and thus made to take up again a portion of carbon; in the other, the carbon is burned out from melted cast-iron by a powerful current of air. The former is known as the process of cementation; Fig. 96 represents a cross-section of the furnace in which it is effected. The iron bars are packed in charcoal in the fire-clay chests or boxes shown in the figure, the whole covered with sand to exclude the air, and heated to redness for from seven to ten days, after which it is allowed to cool down slowly. The bars are now brittle, covered with blisters, and are easily fusible. They may be at once piled together, heated and rolled into bars of "shear-steel;" or they may be broken in small pieces, melted in crucibles of

Fig. 96. Cementation Furnace.

fire-clay, with the addition of a little manganese oxide, and cast into ingots; these ingots are afterward heated and drawn out under the hammer into bars. In this form, it is known as "cast-steel."

The second steel-process is named the Bessemer process, from its inventor. In it, the melted cast-iron is run directly into large wrought-iron vessels lined with fire-clay, called c o n v e r t e r s, Fig. 97, where it meets with a blast of air blown in under a pressure of two entire atmospheres. The iron burns vividly,

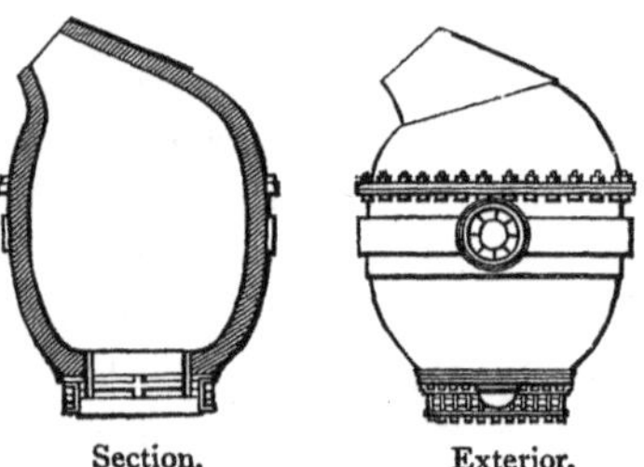

Fig. 97. Bessemer Converter.

the oxide produced acting on the carbon, silicon and other impurities, to oxidize and remove them. At the end of about 20 minutes, the luminous flame suddenly disappears, and a malleable iron containing 0·3 to 0·5 per cent of carbon, is left melted in the converter. To this is now added a suitable proportion of white cast-iron—called technically spiegel-eisen—which contains sufficient carbon to convert the whole into steel. It is then poured into moulds, and the ingots thus obtained are hammered or rolled, as before. Formerly, attempts were made to stop the air-blast at the right point; but the steel thus made varied so widely in quality, that the process above described was substituted. Since the Bessemer process removes the sulphur only partially, and the phosphorus not at all, it is evident that a cast-iron as free from these impurities as possible must be employed.

Steel resembles iron very closely, being distinguished from it only by the remarkable property it possesses of becoming extremely hard and brittle when heated and suddenly cooled. By cautious re-heating, the brittleness thus acquired may be diminished, the steel becoming highly elastic; this process is called tempering. For the manufacture of tools and cutting-instruments, as well as for springs, cementation-steel is preferred; but for many other purposes, as for rails, etc., Bessemer steel is said to be superior to it.

Malleable cast-iron is produced by heating articles made of ordinary white cast-iron to redness for several hours in contact with an iron oxide. A reverse cementation-process takes place, by which the carbon is partially removed and the iron becomes semi-malleable.

IRON AND CHLORINE.

357. Ferric Chloride.—*Formula* Fe_2Cl_6. Ferric chloride has been found native, in crevices in active volcanoes. It is prepared by the direct action of chlorine on iron, or by heating its hydrate. A sublimate of iron-black iridescent scales is obtained, which have a metallic luster, and are volatile a little above 100°, yielding a vapor of density 162·5. Ferric chloride is deliquescent, and dissolves in water with a hissing noise, forming a hydrate. This hydrate is also produced by dissolving iron in hydrochloric acid, heating the solution, and adding nitric acid so long as nitrous gas is evolved. On evaporation, orange-red rhombic crystals are deposited, having the composition Fe_2Cl_6, 6aq.

358. Ferrous Chloride.—*Formula* $FeCl_2$. When chlorine or hydrochloric acid gas is passed over iron-filings in excess, heated to redness, ferrous chloride is obtained in white shining hexagonal scales, which have a specific gravity of 2·5, and are deliquescent in moist air. Ferrous chloride is soluble in two parts of water, and the solution evaporated and cooled away from the air, deposits bluish-green, monoclinic crystals, having the composition $FeCl_2$, 4aq. Ferrous, like ferric chloride, forms double salts with alkali-chlorides.

IRON AND OXYGEN.

359. Ferric Tri-oxide. — *Formula* $Fe^{VI}O_3$. Ferric trioxide, together with the corresponding ferric acid, are both unknown. Certain salts have however been prepared of analogous constitution. Such is potassium ferrate, produced by projecting iron-filings, mixed with twice their weight of niter, into a red-hot crucible. On extracting the mass with ice-cold water, a deep cherry-red solution is obtained, which contains potassium ferrate, but which is very unstable, being easily decomposed into oxygen, potassium hydrate, and ferric oxide. By adding barium chloride to this solution, a purple-red precipitate of barium ferrate is obtained, which is far more stable. It has the composition $BaFeO_4$, aq.

360. Ferric Oxide.—*Formula* $Fe_2{}^{VI}O_3$. This oxide of iron occurs abundantly in nature in two forms; one crystallized in rhombohedrons and mirror-like, called specular iron; the other columnar or fibrous, sometimes amorphous or oolitic in structure and blood-red in color, whence it is called hematite. Artificially, it may be obtained in rhombohedral crystals by igniting the amorphous oxide in a slow current of hydrochloric acid gas; and as an amorphous powder by igniting the hydrate, or ferrous carbonate or oxalate. The latter variety is used for polishing metals and glass, under the name of rouge, crocus, or colcothar. Its specific gravity is about 5. It is reduced to the metallic state when heated in hydrogen. After ignition, it is almost insoluble in acids.

361. Ferric Hydrates exist also in nature, ortho-ferric hydrate, $H_6Fe_2O_6$, in the mineral limnite, mono-metaferric hydrate, $H_4Fe_2O_5$, in xanthosiderite, and di-metaferric hydrate, $H_2Fe_2O_4$, in göthite, beside other and intermediate forms. Ortho-ferric hydrate is precipitated, on adding ammonium hydrate to a solution of ferric chloride, as a bulky brownish-red precipitate, which loses water on drying and forms a yellowish-brown powder. Moist ferric hydrate forms the best antidote in cases of arsenical poisoning. It is used also in calico-printing as a mordant, and in purifying gas.

362. Ferroso-ferric Oxide.—*Formula* $Fe''Fe_2^{VI}O_4$. This oxide of iron occurs native as the mineral magnetite, crystallized in regular octahedrons. The same crystalline form is obtained artificially by fusing ferric phosphate with three or four times its weight of sodium sulphate. This oxide in the amorphous form is the product of the combustion of iron in oxygen. It is a hard, black, metallic-like solid, of specific gravity 5·0 and highly magnetic. It is one of the most valuable of iron ores, containing 72 per cent of this metal. Ferroso-ferric oxide may be regarded as ferrous ferrite, derived from ferric hydrate, $H_2Fe_2^{VI}O_4$; analogous to ferrous chromite $Fe''Cr_2^{VI}O_4$. Other similar compounds are magnesium ferrite, $Mg''Fe_2^{VI}O_4$, known as the mineral magnesioferrite; and zinc ferrite, $Zn''Fe_2^{VI}O_4$, or franklinite, in which, however, a part of the zinc is replaced by iron and manganese. Ferroso-ferric hydrate is precipitated by ammonic hydrate from a solution containing ferrous and ferric salts in suitable proportions, as a brownish-black, dense mass, which dries to a brittle, strongly magnetic powder.

363. Ferrous Oxide.—*Formula* $Fe''O$. Ferrous oxide may be obtained by igniting ferrous oxalate in a close vessel, as a black powder, which takes fire in the air and produces ferric oxide. Ferrous hydrate is precipitated whenever solutions of alkali-hydrates are added to solutions of pure ferrous salts, both being free of air. White flocks are thus produced, which, dried away from the air, have but a slight greenish tinge, but which on exposure, take fire and burn to ferric oxide. It has a strong reducing action.

Ferrous oxide forms a numerous and important class of salts. United to sulphuric oxide, it forms ferrous sulphate $FeSO_4$, known in commerce as green vitriol or copperas. It is produced on the large scale by exposing ferrous sulphide—obtained by roasting ferric disulphide, or pyrite—to the weather, by which it is oxidized :—

$$FeS + (O_2)_2 = Fe''SO_4.$$

It is also the product of the action of sulphuric acid on iron :—

$$H_2SO_4 + Fe = FeSO_4 + H_2.$$

By evaporating its solution, pale green monoclinic prisms crystallize out, having the formula $FeSO_4$, 7aq. The same salt occurs native as the mineral melanterite. The crystals effloresce in dry air, and lose all their water at 300°. They are soluble in one and a half parts of water at 15°, but are insoluble in alcohol. Both the crystals and their solutions readily oxidize in the air. When heated to redness, previously perfectly dried, it gives off sulphurous and sulphuric oxides, leaving a pure ferric oxide. It is therefore used for preparing the Nordhausen or di-sulphuric acid. It is employed in the arts also for dyeing, for tanning, and for making writing-ink.

Ferrous carbonate, $FeCO_3$, occurs native as siderite, or spathic-iron, in obtuse rhombohedrons, light grayish-white in color, and of specific gravity 3·8. It is thrown down, on the addition of a soluble carbonate to a solution of a ferrous salt, as a white precipitate, rapidly passing into brown ferric hydrate on drying. It is soluble in water containing carbonic acid ; occurring native in this form, in chalybeate springs.

IRON AND SULPHUR.

364. Ferric Di-sulphide.—*Formula* $Fe^{IV}S_2$. This sulphide of iron occurs native in two forms; one brass-yellow and isometric, called pyrite; the other white and orthorhombic, called marcasite. Buff suggests for pyrite the formula S═Fe═S, and for marcasite, Fe═S═S. Both varieties, on heating in close vessels, give off sulphur and yield the magnetic sulphide, Fe_3S_4.

365. Ferrous Sulphide.—*Formula* FeS. Ferrous sulphide is produced by the direct union of sulphur and iron, as when iron-wire burns in sulphur-vapor, or when the two substances are melted together in suitable proportions. It is a grayish-yellow solid, with a metallic luster and crystalline structure, and easily fusible. When finely divided, it is oxidized to ferrous sulphate on exposure to the air. With acids it evolves hydrogen sulphide gas. It is precipitated from ferrous solutions by alkaline sulphides, as a hydrate.

§ 4. Nickel and Cobalt.

NICKEL.—*Symbol* Ni. *Atomic weight* 59. *Equivalence* II *and* IV, *probably* VI. *Also a pseudo-triad* $(Ni_2)^{VI}$.

366. History and Occurrence.—Nickel is one of the rarer metals. It was discovered by Cronstedt in 1751 in a copper-colored mineral, to which, having failed in attempting to extract copper from it, the miners had applied in derision, the name kupfernickel. From this mineral, the name nickel is derived. Nickel occurs free in nature only in meteoric irons; in combination, it exists in many minerals, as niccolite (kupfernickel) Ni_2As_2; gersdorffite, $(NiS)''_2As_2$; ullmannite, Ni_2S_2

$(AsSb)_2$; annabergite, $Ni_3(AsO_4)_2$; zaratite, $NiCO_3$ with nickel hydrate; and morenosite, $NiSO_4$, 7aq.

367. Preparation and Properties.—Nickel is generally obtained commercially either from kupfernickel or from an artificial arsenide produced in the manufacture of smalt, and known as speiss. These arsenides are roasted to drive off the arsenic, are then dissolved in hydrochloric acid, the antimony, bismuth, copper, etc., precipitated as sulphides and removed, the iron, after oxidation, precipitated as ferric oxide by ammonia, the ammoniacal solution exposed to the air, and the nickel thrown down by potassium hydrate. The dried precipitate is formed into cubes one or two centimeters on a side, mixed with pulverized charcoal, and placed in intensely heated fire-clay cylinders. The nickel oxide is reduced, though not fused; the small cubes of metallic nickel which are drawn from the bottom of the cylinders being sent in this form, into commerce. Nickel is a pure silver-white, ductile and malleable metal, of specific gravity 8·6. It is exceedingly infusible, and has very great tenacity. It is magnetic, but loses this property at 250°. It tarnishes in moist air, oxidizes readily at a red heat, and is dissolved somewhat slowly by acids. It is largely used for making german silver, which is an alloy of nickel with copper and zinc.

COBALT.—*Symbol* Co. *Atomic weight* 59. *Equivalence* II, IV, *and probably* VI. *Also a pseudo-triad* $(Co_2)^{VI}$.

368. History and Occurrence.—The property of certain cobalt compounds to color glass blue was known to the ancient Greeks and Romans. Its ores were long known to the German miners under the name of cobalt, a term derived from kobold, the evil spirit of the mines, who, as they supposed, tantalizingly offered them an ore rich in appearance, but worthless. The metal was first prepared by Brandt in 1733, and more fully studied by Bergmann in 1780. Cobalt occurs in nature in small quantity in meteorites, but principally in the minerals smaltite or speisskobalt, cobaltite or cobalt-glance, erythrite, or cobalt-bloom, and asbolite, or earthy cobalt.

369. Preparation and Properties.—Cobalt may be obtained by reducing its oxide—obtained from the ammoniacal solution mentioned under nickel—with charcoal, or in a current of hydrogen. It may also be obtained by igniting the oxalate. Cobalt has a steel-gray color, with a tinge of red; it is hard, has a granular fracture, a specific gravity of 8·7 to 8·9, and is malleable at a red heat. It takes up carbon and becomes fusible in an ordinary furnace. It is more magnetic than nickel, and retains its magnetism permanently. When massive, its surface becomes tarnished on exposure to moist air; it oxidizes readily at a red heat, and burns in oxygen. Sulphuric and hydrochloric acids dissolve it slowly with the evolution of hydrogen. Nitric acid acts upon it readily.

370. Compounds of Cobalt.—Cobalt forms both cobaltous and cobaltic compounds. Of these, cobaltous chloride, $CoCl_2$, a blue solid which forms a pink hydrate—hence used for a sympathetic ink—cobaltous nitrate, $Co''(NO_3)_2$, a rose-red salt used as a blowpipe re-agent, cobaltous sulphate, $CoSO_4$, a light-red salt crystallizing with seven molecules of water, and isomorphous with ferrous sulphate; and cobaltic oxide, $Co_2^{VI}O_3$, and the remarkable cobaltamines, roseo-, purpureo-, xantho-, and luteo-cobalt, may here be mentioned. Smalt is an impure cobaltous silicate, made by fusing the roasted ore with potassium carbonate and pulverized quartz. A deep blue glass is thus obtained, which is poured into water and thus finely divided, in which state it is sent into commerce as a pigment. Zaffre is a very impure oxide of cobalt, prepared by roasting the arsenical ores, and mixing the product with twice its weight of sand. It is used for coloring glass blue. Thenard's blue is made by igniting alumina and cobalt phosphate in a covered crucible. Rinman's green is produced by treating mixed zinc and cobaltous oxides in the same way. Both are used as pigments.

EXERCISES.

§ 1.

1. How does chromium occur in nature? Why is it so called?
2. What equivalences has it in its oxides? Its chlorides?
3. Write the graphic formula of mercuric ortho-chromate. Of lead mono-meta-chromate. Of barium di-meta-chromate. Of potassium di-chromate.
4. By the following reaction :—

$$K_2Cr_2O_7 + C_2 = Cr_2O_3 + K_2CO_3 + CO,$$

how much Cr_2O_3 will a kilogram of $K_2Cr_2O_7$ yield?

§ 2.

5. Write the reaction in preparing manganese.

6. What weight of potassium permanganate may be obtained from a kilogram of MnO_2, the reaction being,—

$$(MnO_2)_2 + O_3 + K_2O = K_2Mn_2O_8 \, ?$$

7. Calculate the percentage composition of manganous silicate.

§ 3.

8. What per cent of iron do the four ores mentioned, contain?
9. How is cast-iron obtained? What are its varieties?
10. What is the weight of a cubic dekameter of gray-iron?
11. What are the chemical changes in the refining of iron?
12. Give the chemistry of the two processes for steel.
13. In what do cast-iron, wrought-iron and steel differ?
14. 250 grams pure iron are burned in an excess of chlorine; What compound is formed? What weight of it?
15. Answer the above questions, using oxygen in place of chlorine.
16. What percentage of water is there in the mineral göthite?
17. 1,000 kilograms pyrite is roasted and exposed to the weather; what weight of crystallized ferrous sulphate may be obtained by its oxidation?
18. One c.c. spathic iron contains what volume of CO_2?

§ 4.

19. What percentage of nickel does niccolite contain?
20. An iron vessel weighs 156 grams; what will be its weight if made of nickel?
21. By what chemical process is nickel obtained?
22. A cube of cobalt weighs 50 grams; what does it measure on a side?
23. What weight of cobaltous oxide is required to yield 36 grams of cobaltous nitrate? Of crystallized sulphate?

CHAPTER EIGHTH.

POSITIVE TETRADS.

§ 1. LEAD.

Symbol Pb. *Atomic weight* 207. *Equivalence* II *and* IV.

371. History and Occurrence.—Lead has been known from the earliest ages of history. It is mentioned in the book of Job, and elsewhere in the sacred writings. The Romans worked the lead-ores of Spain and of England, and the Carthaginians those of Spain, the extent of their mining and smelting operations exciting surprise, even at the present day. The principal workable ore of lead is its sulphide, or galenite; though it occurs also somewhat abundantly as carbonate, or cerussite; as sulphate, or anglesite; as chloro-arsenate, or mimetite; as chloro-phosphate or pyromorphite; and in other forms.

372. Preparation and Properties.—Lead is prepared from the sulphide by a comparatively simple metallurgical process. The ore is first roasted on the floor of a reverberatory furnace, by which both oxide and sulphate of lead are produced. The furnace is then closed tight, and these products react upon the undecomposed lead sulphide as follows:—

$$\underset{\text{Lead oxide.}}{(PbO)_2} + \underset{\text{Lead sulphide.}}{PbS} = \underset{\text{Sulphurous oxide.}}{SO_2} + \underset{\text{Lead.}}{Pb_3},$$

$$\underset{\textit{Lead sulphate.}}{PbSO_4} + \underset{\textit{Lead sulphide.}}{PbS} = \underset{\textit{Sulphurous oxide.}}{(SO_2)_2} + \underset{\textit{Lead.}}{Pb_2}.$$

When the ore contains impurities, it is smelted by fusing it with iron, ferrous sulphide and lead being produced. Other methods are employed for smelting lead, varying according to the ore and the locality.

Lead is a brilliant metal, bluish-white in color, and so soft as to be easily cut with a knife. It leaves upon paper a bluish-gray streak and is very malleable. It has a specific gravity of 11·4, crystallizes in regular octahedrons, and fuses at 325°. At a red-heat, it is slightly volatile. It has but a feeble tenacity, a wire two millimeters in diameter sustaining only nine kilograms. A freshly-cut surface tarnishes in ordinary air, but remains bright in perfectly dry air and also in water free from air. Potable waters in general act upon lead, dissolving it, and partly precipitating it as carbonate. This action is particularly noticeable in well-waters, which contain nitrates, from decomposed animal matter, or chlorides, from saline infiltration. Lead water-pipes should therefore be avoided or used with great caution. When melted, lead is rapidly converted into the oxide. It is scarcely attacked by sulphuric or hydrochloric acid at ordinary temperatures, but dissolves readily in nitric acid. In presence of air and moisture, it is acted upon by quite feeble acids as acetic and carbonic. Hence, the use of such vessels as are made of, or united with, lead, or lead solder, should be avoided for such articles. Lead when taken into the system, unites definitely with certain tissues and is retained there, until finally, sufficient accumulates to produce poisoning. Acute colic is characteristic of poisoning by a large dose of lead; but in chronic poisoning, which is far more common, there is paralysis, particularly of the muscles of the fore-arm, causing the wrists to drop ; or there may be simply an indefinable feeling of malaise, accompanied by dyspeptic symptoms.

Lead is used extensively in the arts for various purposes, both alone and alloyed with other metals. With a small proportion of arsenic, it forms the alloy of which shot are made ; with antimony and tin, it forms type-metal ; with bismuth, the soft alloy used for permanent pencil-points ; with tin, pewter and soft solder ; and with cadmium, tin, and bismuth, fusible metal melting at 60°.

EXPERIMENT.—Metallic lead in a state of fine division, takes fire readily in the air, forming what is known as a pyrophorus. To obtain it in this form, tartrate of lead is produced by adding lead acetate to a solution of potassio-sodium tartrate,—rochelle salt,—so long as it forms a precipitate. The lead tartrate, filtered off, washed, and dried, is placed in a tube of hard glass, the tube is then drawn out at the end, and the whole is heated to bright redness, until no more fumes escape. The end of the tube is then sealed, and the whole allowed to cool. On breaking the tube and pouring out the contents into the air, the metallic lead at once inflames, producing a shower of fire. In oxygen, the combustion is brilliant ; in carbonic di-oxide, no combustion takes place.

LEAD AND CHLORINE.

373. Plumbic Chloride.—*Formula* $Pb''Cl_2$. This chloride has been found in the crater of Vesuvius after an eruption, and is known as cotunnite. It is precipitated from any plumbic solution, if sufficiently concentrated, on the addition of hydrochloric acid or a chloride. It is a heavy white powder, soluble in 135 parts of cold and 30 parts of boiling water, from which it crystallizes on cooling, in lustrous needles. It melts when heated in close vessels, and at a higher temperature, sublimes. The fused chloride is translucent and sectile, and is known as horn-lead. A white and a yellow oxy-chloride are used as pigments.

374. Plumbic Per-chloride.—*Formula* $Pb^{IV}Cl_4$. Plumbic per-chloride is obtained in solution by dissolving plumbic peroxide in cold hydrochloric acid, and in crystals by evaporating this solution in vacuo. It has been but little studied.

LEAD AND OXYGEN.

375. Plumbic Peroxide.—*Formula* $Pb^{IV}O_2$. Plumbic peroxide is best prepared by precipitating a solution of four parts of lead acetate with a solution of three and a half parts of crystallized sodium carbonate, and passing chlorine gas through the mixture until the whole of the white carbonate of lead is converted into the brown peroxide. It forms on drying a chocolate-brown, or puce-colored powder, which gives off oxygen on heating. It is a strongly oxidizing agent, uniting directly with sulphurous oxide to form plumbic sulphate. Digested with ammonic hydrate, it forms water and lead nitrate; mixed with one-fifth its weight of sulphur, it inflames spontaneously; rubbed in a mortar with one-sixth of grape-sugar, ignition takes place; it sets iodine free from potassium iodide, and bleaches a solution of sulph-indigotic acid.

Plumbic peroxide combines directly with potassium, sodium, calcium, and even plumbic oxides, forming salts called plumbates, having the general formula M_2PbO_3. Potassium plumbate occurs in white octahedrons, decomposible by water. Plumbic plumbates form the various bodies known as red-leads. The compound $Pb''_2Pb^{IV}O_4$, or plumbic ortho-plumbate, written more often Pb_3O_4, occurs native as minium. This, as well as the meta-plumbate $Pb''Pb^{IV}O_3$, or Pb_2O_3, is produced largely in the arts as a pigment, by oxidizing litharge in a current of air, and then cooling very slowly. Nitric acid decomposes these plumbates, producing lead nitrate and plumbic peroxide.

376. Plumbic Oxide.—*Formula* $Pb''O$. This oxide of lead occurs native as the mineral massicot. It is prepared on the large scale in the arts under the name of litharge, by heating melted lead in a current of air. Its color is pale-yellow, or orange-yellow, according to the temperature at which it is produced. It is dimorphous, crystallizing in rhombic octahedrons and in regular dodecahedrons. Its specific gravity is 9·42, and it fuses at a full red-heat. It is soluble only in 7,000 parts of water, but acids dissolve it easily, forming definite salts. It is also soluble in alkali-hydrate solutions, and in lime-water. It is used in the arts in the manufacture of glass.

Plumbic hydrate, H_2PbO_2 or $Pb''(OH)_2$, is known only as a colorless, sweetish, alkaline liquid, obtained by the action upon lead of water and air, free from carbon di-oxide. The precipitate produced by hydrates of the alkalies in plumbic solutions, is an oxyhydrate having the composition $Pb_2O(OH)_2$.

Plumbic nitrate, $Pb''(NO_3)_2$, is produced by dissolving lead or its oxide in nitric acid and crystallizing. The hydro-nitrate $H(NO_2)'PbO_2$ or $Pb\begin{cases}OH\\O(NO_2)\end{cases}$, is precipitated from the nitrate, by adding ammonium hydrate in small quantity. Plumbic carbonate,

$PbCO_3$, occurs native as cerussite; a hydro-carbonate, $Pb_3(CO_3)_2(OH)_2$, is much used as a pigment under the name of white-lead. Plumbic sulphate, $PbSO_4$, which occurs native as anglesite, is precipitated from solutions of lead on adding sulphuric acid or soluble sulphates.

397. Plumbous Oxide.—*Formula* Pb_2O. When lead oxalate is heated to 300° in a closed vessel, a black velvety powder is obtained which is plumbous oxide. It contains no lead, and no plumbic oxide; but takes fire when heated in the air, producing this oxide.

INDIUM. — *Symbol* In. *Atomic weight* 74. *Equivalence* II.

378. History, Preparation, and Properties. — Indium was discovered in the Freiberg zinc-blendes, by Reich and Richter, in 1863. They observed that the impure zinc chloride obtained by dissolving the roasted ore in hydrochloric acid, contained a substance whose spectrum consisted of two indigo-blue lines, one of which was very brilliant and more refrangible than the other. Indium is a soft, malleable metal, resembling lead, of specific gravity 7·2. It is unaltered in the air, and leaves a gray streak on paper. At a red heat it melts, and burns with a violet light, producing the oxide. Hydrochloric and sulphuric acids dissolve it readily. Indium oxide, InO, is a straw-yellow powder, becoming brown on heating, and easily reducible on charcoal. Indium hydrate, $In''(OH)_2$, is thrown down, as a white precipitate, from solutions of indium by alkali-hydrates. Indium sulphide, InS, is a dark yellow powder. Indium chloride, $InCl_2$, is procurable as a white crystalline sublimate which is deliquescent.

§ 2. Platinum.

Symbol Pt. *Atomic weight* 197. *Equivalence* II *and* IV.

379. History and Occurrence.—Platinum was brought to Europe from South America in 1735 by Ulloa, and in 1741 by Wood. It was first described by Watson in 1750; and independently, by Scheffer in 1752. It derives its name from the word platina, the Spanish diminutive of plata, silver. It occurs native usually in rounded grains, though sometimes it is found crystallized in octahedrons. It is rarely pure, the native platinum containing gold, iron, and copper, beside iridium, ruthenium, osmium, rhodium and palladium, its natural congeners. It occurs not only in South America, but also in Russia, in Borneo, and in California.

380. Preparation and Properties.—Native platinum was formerly purified by the method of Wollaston, which consists in treating the crude metal, first with nitric acid and then with hydrochloric acid, and afterward with boiling aqua regia. By the latter treatment, the platinum, palladium, and a portion of the rhodium, are dissolved, while a mixture of iridium, rhodium, osmium, and ruthenium, known as iridosmine, is left. The platinum is thrown down from its solution by ammonium chloride, as ammonium chloro-platinate. This, on ignition, leaves the metal in a finely-divided state known as spongy platinum, which is condensed into a cake by powerful pressure, and is then welded at a white heat, into a homogeneous mass.

Latterly, however, Deville's method has almost

entirely superseded that of Wollaston. In this the crude platinum is melted with an equal weight of lead sulphide, and half its weight of metallic lead; in this way, the platinum is dissolved, leaving the iridosmine. The platinum-lead alloy is then melted and exposed to a current of air, by which the lead is oxidized, the oxide flowing off as slag, and the platinum being left as a porous mass. This is then placed in a furnace made of lime, Fig. 98, and by means of two powerful oxy-hydrogen jets, it is melted and cast into ingots. Masses weighing 100 kilograms have been produced by this process at one fusion. The melted mass absorbs oxygen, and evolves it again on cooling, like silver.

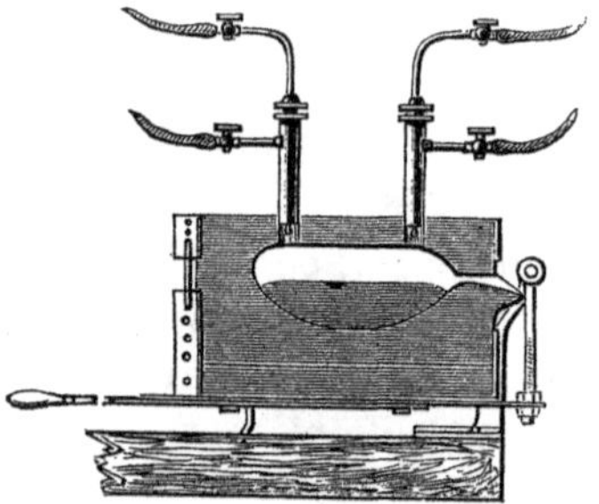

Fig. 98. Platinum Furnace.

Platinum is a brilliant white metal, with a tinge of blue. It has a specific gravity of 21·5, is extremely malleable and ductile, and has a tenacity and hardness resembling that of copper. It is an imperfect conductor of heat and of electricity, and is infusible by ordinary means, but yields to the oxy-hydrogen flame, in which it is partially volatilized. Before complete fusion, it softens, and may then be welded. At high temperatures it absorbs hydrogen, and is readily permeable by this gas at a red heat. Platinum is unaltered in the air at any temperature; it is not attacked by any single acid, being dissolved only by aqua regia. Fused potassium and sodium hydrates act upon it; and it combines directly with sulphur, phosphorus, arsenic, and silicon.

Platinum possesses in a remarkable degree the property of condensing gases upon its surface. In the form of platinum foil, it will cause the explosion of mixed oxygen and hydrogen gases; but in the form of platinum-sponge, it is much more active; and in the still more finely divided form known as platinum black, it is capable of absorbing 800 times its volume of oxygen. Platinum black, therefore, owing to this condensed oxygen, is an energetic oxidizing agent; alcohol thrown upon it is at once inflamed.

Owing to its unalterability by chemical agents, platinum is used extensively both in the arts and in the laboratory for chemical vessels. Large platinum stills for sulphuric acid, weigh often 30,000 grams. Platinum has been also used in Russia, for coinage.

COMPOUNDS OF PLATINUM.

381. Platinum Chlorides.—Two chlorides of platinum are known, the platinic chloride, $PtCl_4$, and the platinous chloride, $PtCl_2$. The former is obtained whenever platinum is dissolved in aqua regia. By evaporation at 100°, a brown-red deliquescent mass is left, soluble freely in water, alcohol and ether. It loses half its chlorine at 230°, and the whole at a red heat. It unites directly with alkali chlorides, forming chloro-platinates, M_2PtCl_6. Platinous chloride is produced by heating platinic chloride to 230°, till chlorine ceases to be evolved. A dark-green powder is left, insoluble in water, sulphuric and nitric acids, but soluble in sodium and potassium hydrates. A series of remarkable compounds is formed by the action of ammonia upon this chloride, called the ammonio-platinum bases.

382. Platinic Oxide, PtO_2, forms a hydrate $Pt^{IV}(OH)_4$, which dissolves in alkali-hydrates yielding platinites. Platinous oxide, PtO, yields a basic hydrate $Pt''(OH)_2$, which forms salts with acids. Platinous and platinic sulphides are also known.

§ 3. ALUMINUM.

Symbol Al. *Atomic weight* 27·5. *Equivalence* $(Al_2)^{VI}$.

383. History and Occurrence.—Aluminum oxide, or alumina, was long confounded with lime, from which it was first distinguished by Marggraff in 1754. Oersted, in 1826, first prepared the chloride; and from this, in 1828, Wöhler obtained the metal. His process was made a commercial one by St. Claire Deville in 1854. The metal was first prepared from cryolite by H. Rose in 1855.

Aluminum, next to oxygen and silicon, is the most abundant element in nature. The mineral corundum is aluminum oxide; diaspore, chrysoberyl, and spinel, are aluminates; micas, feldspars, and clays, are aluminum silicates; and many other minerals contain it as an essential constituent. Its name comes from the Latin alumen, alum, which substance was largely imported into Europe from the east until the fifteenth century.

384. Preparation and Properties.—Aluminum is prepared commercially by the process of Deville, which consists in reducing the chloride with sodium. At the works of Morin in Paris, ten parts sodio-aluminum chloride, five parts of fluor-spar or cryolite, and two parts of sodium, are mixed together and thrown upon the hearth of a reverberatory furnace, previously heated to full redness. A violent action takes place, great heat is evolved, and the liquefied mass of slag and metal collects at the back of the furnace. The latter is drawn off and cast into ingots. As much as eight kilograms are obtained at one operation. Tissier at Amfreville makes aluminum from the mineral cryolite, after the method proposed by H. Rose.

Aluminum is a brilliant bluish-white metal, capable of taking a fine polish. It crystallizes in octahedrons, and conducts heat and electricity readily. It is highly sonorous, is very light, specific gravity 2·56 to 2·67, and is very malleable and ductile. Its tenacity is about equal to that of silver and it is slightly magnetic. After fusion it is soft, but becomes hard by hammering. It melts at a temperature but little above the fusing-point of zinc, but is not volatile. It tarnishes very slowly in the air, though it burns readily in thin leaves, when heated in oxygen. Sulphuric and nitric acids are without action upon it, though hydrochloric acid and alkali-hydrates attack it readily.

Aluminum is used in the arts chiefly for ornamental purposes, for which its luster, its whiteness, and its unalterability in the air, well adapt it. Its lightness makes it useful for weights, and for astronomical and physical instruments. Culinary utensils have also been made of it. Alloyed with copper, it forms aluminum-bronze, which, when it contains ten per cent of aluminum to ninety of copper, is hard, malleable, as tenacious as steel, and takes a fine polish.

COMPOUNDS OF ALUMINUM.

385. Aluminum Chloride. — *Formula* $Al_2^{VI}Cl_6$. The only known compounds of aluminum contain this element with a double atom, two tetrad atoms thus forming a hexad. Did not its molecular heat forbid, it would be proper to consider this double atom as one, with an atomic weight of 55. Aluminum chloride is prepared by passing chlorine over an ignited mixture of the oxide and charcoal. It is a colorless, semi-crystalline, waxy substance, fusible and volatile, and having a vapor-density of 134. It combines violently with water, forming the hydrate Al_2Cl_6, 12 aq. Chlorine, passed over an ignited mixture of salt, aluminum oxide, and charcoal, yields sodio-aluminum chloride, $Na_2Al_2Cl_8$, as a white crystalline solid, which melts at 200°. Aluminum fluoride, Al_2F_6, occurs native in the mineral cryolite, combined with sodium fluoride.

386. Aluminum Oxide.—*Formula* Al_2O_3. Aluminum oxide occurs native in the mineral corundum, which includes the precious stones known as the ruby and the sapphire, as well as the valuable polishing material called emery. It may be prepared by the combustion of the metal in oxygen, or by igniting the hydrate. It is found in nature crystallized in rhombohedral forms, with a specific gravity of 3·9, and scarcely inferior in hardness to the diamond. Aluminum hydrate, $Al_2(OH)_6$, or ortho-aluminic base, occurs native as gibbsite; it is obtained by precipitating any salt of aluminum by ammonium hydrate. Mono-meta-aluminic base $Al_2O(OH)_4$, and di-meta-aluminic base $Al_2O_2(OH)_2$, or $H_2Al_2O_4$, are also known, the former occurring in nature as the mineral beauxite, the latter as diaspore. Aluminic hydrate acts as an acid with basic radicals, forming aluminates; sodium ortho-aluminate, $Na_6Al_2O_6$, and meta-aluminate, $Na_2Al_2O_4$, and potassium meta-aluminate, $K_2Al_2O_4$, have been prepared artificially; and glucinum aluminate, $G''Al_2O_4$, or chrysoberyl; magnesium aluminate, $Mg''Al_2O_4$, or spinel; and zinc aluminate, $Zn''Al_2O_4$, or gahnite, are well-known minerals. With strong acids, aluminum hydrate acts as a base. With sulphuric acid, it yields aluminum sulphate, $Al_2(SO_4)_3$, 18 aq., the mineral alunogen; it is used in dyeing. It forms a characteristic class of double sulphates called alums, whose general formula is $M_2Al_2(SO_4)_4$, 24 aq., M being generally K or (NH_4). They all crystallize in regular octahedrons, are acid in their reaction and have a styptic taste. Aluminum phosphate, $Al_2(PO_4)_2$, is known in the hydrated form. The silicates of aluminum constitute an important class of minerals. Cyanite is meta-aluminum silicate, $SiAl_2{}^{VI}O_5$. Orthoclase is a potassio-aluminum silicate, albite a sodio-aluminum silicate, and labradorite a calcio-aluminum silicate. Clay is a more or less pure aluminum silicate, $Al_2Si_2O_7$, 2 aq. It is used in pottery, the finer kinds being known as kaolin, or porcelain clay.

EXERCISES.

§ 1.

1. Mention the minerals in which lead occurs.
2. Write the chemical reactions which take place in obtaining lead from galenite.
3. 1000 kilograms of galenite yield what volume of SO_2?
4. What weight of metallic lead would be obtained?
5. Calculate the volume occupied by a kilogram of lead.

6. What are the objections to the use of lead water-pipes?

7. How is a lead pyrophorus prepared?

8. What volume of chlorine is contained in the plumbic chloride required to saturate a liter of water?

9. What weight of sulphur may be burned by one gram of lead di-oxide? Write the reaction.

10. Give the rational constitution of the red-leads.

11. One cubic decimeter of lead will give what weight of litharge?

12. How many kilograms of white lead can be obtained from 1000 kilograms of lead?

13. How was INDIUM discovered? From what was it named?

§ 2.

14. With what other elements is platinum associated?

15. By what two methods is it refined?

16. An alloy of platinum and gold has a specific gravity of 20; what per cent of platinum does it contain?

17. What remarkable property has finely-divided platinum?

18. What pressure would condense oxygen as it is condensed by platinum black?

19. One gram of $PtCl_4$ contains what volume of chlorine?

§ 3.

20. In what compounds does aluminum occur in nature?

21. Give the process adopted by Morin for preparing it.

22. Cryolite has the formula $Na_6Al_2F_{12}$; what per cent of aluminum does it contain?

23. What is the weight of an aluminum ball, 5 centimeters in diameter?

24. Write the graphic formula of aluminum chloride. Oxide.

25. Under what names does aluminum oxide occur?

26. Calculate the percentage composition of the mineral chrysoberyl.

27. Fibrolite contains 36·8 per cent of silicic oxide and 63·2 per cent of aluminum oxide; what is its formula? mol. wt. 16

CHAPTER NINTH.

POSITIVE TRIADS.

§ 1. GOLD.

Symbol Au. *Atomic weight* 196·6. *Equivalence* I *and* III.

387. Occurrence.—Gold occurs quite widely distributed in nature, though in small quantity. It is found generally in quartz veins intersecting metamorphic rocks of various ages, and in the alluvial detritus which has resulted from the disintegration of these rocks. To extract the gold from the auriferous quartz, the whole is finely pulverized, and then treated with mercury, which dissolves the gold. By pressing out the excess of mercury and distilling off that which is left in the solid amalgam, the gold is obtained pure. The yield of the United States gold-fields in 1866 was $86,000,000, and that of Australia $30,000,000. The largest single mass was found at Ballarat, Australia; it weighed 184⅔ pounds.

388. Properties.—Gold is a soft, orange-yellow metal, of specific gravity 19·33, and very brilliant. It is very ductile, and so malleable that it may be beaten into leaves only one ten-thousandth of a millimeter thick. These gold-leaves transmit green light, though when rendered non-lustrous by heat, this light is ruby-red. Gold crystallizes in isometric forms, conducts heat and electricity well, and fuses at 1250°. It is unaltered in the air, and is not attacked by any single acid or alkali-hydrate, though solutions which contain free chlorine, like aqua regia, dissolve it readily.

Gold is used both for jewelry and for coinage. Being too soft for either purpose alone, it is alloyed with either copper or silver, the mint-alloy of the United States consisting of nine parts of gold and one of copper. The purity of gold for jewelry is estimated by the carat, pure gold being twenty-four carats fine; hence an alloy of eighteen parts gold to six of silver and copper, is said to be eighteen carats fine. Gold is freed from copper by cupellation, and from silver by q u a r t a t i o n. The latter process consists in adding silver to the alloy until the gold forms one-quarter part, and then dissolving out the silver by nitric acid. The gold is left pure.

EXPERIMENT.—Prepare a dilute solution of gold by adding to a liter of water a few drops of a rather strong solution of auric chloride; then drop into it one or two fragments of phosphorus of the size of a mustard-seed, and place the whole in the sunlight. Soon, often in the course of a few hours, the water will have a distinct purplish tint. This will deepen in color until finally, if the solution has the proper strength, a beautiful ruby-red liquid will be obtained. Faraday proved that the color of this liquid is due to finely-divided *metallic gold*; so finely divided indeed, that, though gold is one of the heaviest of metals, it does not settle out of the liquid on standing.

COMPOUNDS OF GOLD.

389. Auric Chloride.—*Formula* $AuCl_3$. This chloride is produced by dissolving gold in aqua regia and removing the excess of hydrochloric acid by evaporation to dryness. A dark-red, crystalline, deliquescent mass is obtained, which is freely soluble in water, in alcohol, and in ether. It forms yellow double chlorides with hydrogen, sodium, potassium, and ammonium chlorides.

Aurous chloride, AuCl, is obtained by gently heating auric chloride. It is a nearly white mass, permanent in the air, and insoluble in water. By boiling water, it is decomposed into auric chloride and metallic gold.

390. Auric Oxide, Au_2O_3, is obtained as a dark-brown powder by digesting magnesia in a solution of auric chloride, decomposing the magnesium aurate by nitric acid, and drying the residue at 100°. It is decomposed by light and heat, and unites readily with positive oxides to yield aurates, having the general formula $M'AuO_2$. Ammonium aurate, known as fulminating gold, is violently explosive. In potassio-auric sulphite, $K_3Au'''(SO_3)_3$, auric oxide is basic or positive. Aurous oxide, Au_2O, is a greenish powder; one of its salts, sodio-aurous sulpho-sulphate, $Na_3Au'(S_2O_3)_2$, 2 aq., is used in photography.

§ 2. THALLIUM.

Symbol Tl. *Atomic weight* 204. *Equivalence* I *and* III.

391. History and Occurrence. — Thallium was discovered in 1861, by Crookes, in a residue left after the distillation of selenium from a lead-chamber deposit obtained from a sulphuric acid factory in the Harz mountains. This residue colored a gas-flame intensely green and gave a spectrum consisting of one brilliant green line. From the peculiar tint of this line, Crookes named the metal thallium, from *θαλλός*, a green shoot. Thallium was independently discovered by Lamy, nearly a year later.

By the aid of the spectroscope, which will detect one-four-thousandth of a milligram of thallium, this element has been proved to be widely distributed in nature, though in very minute quantities. The rare mineral crookesite is a selenide of copper, thallium, and silver, and contains seventeen per cent of thallium. Various specimens of iron and copper pyrites contain from a four-thousandth to a hundred-thousandth of their volume of thallium; it occurs also in zinc-blende, in certain varieties of lepidolite, and in the saline residues at Nauheim.

392. Preparation and Properties.—Thallium is generally obtained from the dust deposited in the tubes which, in sulphuric acid works, convey the sulphurous oxide from the pyrites burners to the leaden chambers. This is extracted with water, and the thallium is thrown down from this solution as chloride. This, by solution in sulphuric acid, gives the sulphate; and from this, metallic thallium may be precipitated by zinc.

Thallium is a brilliant, nearly white metal, softer than lead, and having a specific gravity of 11·8. It is malleable but not ductile, owing to its want of tenacity. It melts at 294°, and crystallizes in octahedrons. It tarnishes rapidly in air and burns brilliantly in oxygen. Nitric acid dissolves it readily. It is strongly dia-magnetic.

393. Thallic Chloride, $TlCl_3$, is obtained by acting on thallium with an excess of chlorine. It forms double salts with the chlorides of the alkali-metals. Thallous chloride, $TlCl$, is precipitated from thallous solutions by chlorides; it is white and crystalline, resembling lead chloride. Thallic oxide, Tl_2O_3, is a dark-red powder. Thallous oxide, Tl_2O, is reddish-black in color, and dissolves in water forming a hydrate. With acids it yields salts, thallous sulphate being $Tl_2(SO_4)$.

CHAPTER TENTH.

POSITIVE DYADS.

§ 1. COPPER.

Symbol Cu. *Atomic weight* 63·5. *Equivalence* $(Cu_2)''$ *and* II.

394. History and Occurrence.—Copper has been known from the earliest times. The Romans obtained it from the island of Cyprus and called it aes cyprium, a term which afterward became cuprum, from which the English word copper is derived. Copper is found abundantly in nature both free and in combination. Native copper occurs in masses of great size near Kewenaw Point, Lake Superior, one of which weighed over 400 tons. The workable ores of copper are the oxides cuprite and melaconite; the sulphides chalcocite, covellite, bornite, and chalcopyrite; the sulph-antimonite tetrahedrite; and the carbonates malachite and azurite.

395. Preparation and Properties.—The method applied for the extraction of copper varies with the ore under treatment. The oxides and carbonates are simply heated with charcoal or other fuel, with the addition of some siliceous flux. The sulphides are first roasted, by which the iron sulphide becomes oxide, and some copper oxide is produced. The mass is then fused with silica, thus forming a slag containing the iron as silicate, while a purer copper sulphide is left. By repeating this roasting and the subsequent fusion several times, a nearly pure copper sulphide is obtained. This is again roasted, the copper oxide now produced acting upon the copper sulphide to give metallic copper, thus:—

$$Cu_2S + (CuO)_2 = SO_2 + Cu_4.$$

The metal is refined by being kept melted for many hours with free access of air. The more oxidable metals, together with some copper, pass into the slag as oxides, while some of the copper oxide dissolves in the melted metal. To remove this, it is stirred with the trunk of a young tree, by which this oxide is reduced, the copper becoming at the same time tough and fibrous. If this "poling" be continued too long, the metal is made brittle again. Perfectly pure copper may be obtained by electro-deposition, or by reducing the oxide by a current of hydrogen gas.

Copper is a lustrous metal, flesh-red in color, and somewhat softer than iron. It crystallizes in isometric forms, has a specific gravity of 8·95, and conducts heat and electricity readily. It may be drawn into fine wire, or beaten into thin leaves. Its tenacity is considerable, a wire two millimeters in diameter sustaining a weight of 140 kilograms. It melts at about 1200°, and is volatile at very high temperatures. It is unaltered in ordinary air when massive; though when finely divided, it often takes fire spontaneously. When heated to redness, scales of oxide form on its surface. It is attacked readily by chlorine and sulphur, and by nitric acid. Weak acids and alkalies, and saline solutions act on it slowly; hence, as all its salts are poisonous, anything to be taken as food, should not be prepared in vessels made of this metal or any of its alloys.

Copper finds extensive use in the arts, both as such and in the form of alloys. The alloys of copper with tin and aluminum have been mentioned. With zinc, it forms the well-known alloy brass, of which yellow- or sheathing-metal, and aich-metal, are varieties. Sterro-metal is a brass containing nearly one per cent of tin and two per cent of iron. German silver is an alloy of copper, zinc, and nickel.

COMPOUNDS OF COPPER.

396. Cupric Chloride, $CuCl_2$, is obtained by the action of chlorine upon copper, or by drying its hydrate at 200°. It is a yellowish-brown deliquescent powder, soluble in water, forming a green solution which on evaporation deposits crystals of the hydrate, $CuCl_2$, 2 aq. It forms double salts with the alkali-chlorides. Cuprous chloride, Cu_2Cl_2 or $\begin{cases} CuCl \\ CuCl \end{cases}$, is formed by the action of metallic copper upon cupric chloride. On pouring the solution thus obtained into water, a dense, white, crystalline precipitate is thrown down, which becomes blue on exposure to air, and is fusible at a red heat. It also forms double salts with the chlorides of the alkali-metals. A cuprous hydride, Cu_2H_2, is known.

397. Cupric Oxide, CuO, occurs native as melaconite. It may be prepared by heating the metal in the air, or by calcining the hydrate, carbonate, or nitrate. It occurs in isometric forms—perhaps also in orthorhombic—but is generally massive. Its specific gravity is 6·3 and it fuses without change at a bright-red heat. It is easily reduced when heated with combustible substances, and hence is used in organic analysis. Cupric hydrate, $Cu(OH)_2$, is thrown down as a pale-blue precipitate, on adding sodium hydrate to a cold solution of a cupric salt. It acts strongly basic, and forms numerous salts. Cupric nitrate, $Cu(NO_3)_2$, is produced when copper is dissolved in nitric acid. On evaporation, bright-blue crystals separate, which have the formula $Cu(NO_3)_2$, 3 aq. Cupric sulphate, $CuSO_4$, commonly known as blue vitriol, is obtained by dissolving the oxide, carbonate, or hydrate, in sulphuric acid, or by roasting the sulphide. It crystallizes with five molecules of water, in triclinic prisms, which are soluble in three and one-half times their weight of cold water. Cupric ortho-phosphate, HCu_2PO_5, occurs native as the mineral libethenite. Two cupric ortho-carbonates occur native: malachite, which is green, Cu_2CO_4, aq., and azurite, which is blue, $H_2Cu_3(CO_4)_2$. Cuprous oxide, Cu_2O, forms the mineral cuprite; it is obtained by reducing a solution of copper with grape-sugar in presence of an alkali-hydrate. It is a bright-red powder, isometric in its natural forms, and of specific gravity 6·0. Acids decompose it into cupric oxide and copper.

398. Cupric Sulphide, CuS, is found native as the mineral covellite; it is hexagonal in its crystallization, is of a bluish-black color, and has a specific gravity of 4·6. It is precipitated from cupric solutions by hydrosulphuric acid. Cuprous sulphide, Cu_2S, also occurs native, forming the mineral chalcocite. It is produced abundantly in smelting copper. It crystallizes in orthorhombic prisms, in blackish lead-gray in color, and is easily fusible. It acts as a sulphur base, forming cuprous sulph-arsenites and antimonites.

§ 2. MERCURY.

Symbol Hg. *Atomic weight* 200. *Equivalence* $(Hg_2)''$ *and* II. *Density* 100. *Molecular weight* 200. *Molecular volume* 2. *One liter of mercury-vapor weighs* 8·96 *grams (* 100 *criths).*

399. History and Occurrence.—Mercury has been known from the earliest times; it is mentioned by Theophrastus, as *χυτὸς ἄργυρος*, whence the Latin argentum vivum, and the English quicksilver. The name mercury was given by the alchemists, from the planet of that name, to all volatile substances; but only this one has retained it. Mercury occurs native in small quantity; but the chief ore of mercury is the sulphide, called cinnabar, which is found principally in Idria in Austria, Almaden in Spain, and New Almaden in California.

400. Preparation.—In Austria and Spain, the mercury is obtained by roasting the ore, and carrying the volatile products through a series of chambers, or of long narrow pipes, in which the metal is condensed. A better process is to distill the sulphide with lime or iron-scales in close vessels.

401. Properties.—Mercury is a brilliant, silver-white metal, of specific gravity 13·59. At ordinary temperatures it is a coherent liquid, but cooled to $-40°$ it solidifies to a malleable tin-white mass, easily sectile, and crystallizing in regular octahedrons. It is slightly volatile even at 15°; it boils at 350°, yielding a colorless vapor of specific gravity 6·976. Mercury is unalterable in the air, but at its boiling-point it slowly absorbs oxygen, a crystalline, dark-red oxide forming on the surface. Hydrochloric and dilute sulphuric acids do not attack it, but boiling strong sulphuric acid, and even dilute nitric acid, dissolve it readily. Chlorine and sulphur unite directly with it.

Mercury is used in the arts for filling thermometers and barometers, and very extensively for extracting gold and silver from their ores. Its alloys with other metals are called amalgams. Tin-amalgam is used to cover mirrors.

COMPOUNDS OF MERCURY.

402. Mercuric Chloride.—*Formula* $HgCl_2$. This chloride, often called corrosive sublimate, is formed when mercury is acted on by an excess of chlorine. It is usually prepared by subliming a mixture of mercuric sulphate and salt:—

$$\underset{\textit{Mercuric sulphate.}}{Hg(SO_4)} + \underset{\textit{Sodium chloride.}}{(NaCl)_2} = \underset{\textit{Sodium sulphate.}}{Na_2(SO_4)} + \underset{\textit{Mercuric chloride.}}{HgCl_2}.$$

A white, semi-transparent, crystalline mass is thus obtained, which has a specific gravity of 5·43, melts at 265°, and boils at 295°, yielding a vapor of normal density. It is soluble in sixteen parts of cold and three parts of hot water; from the latter solution it crystallizes on cooling, in long white orthorhombic prisms. Its solution tastes acrid, nauseous, and metallic, and reacts acid; it is an energetic corrosive poison. It coagulates albumin, forming an insoluble compound with it; whence white of egg is the best antidote for it in cases of poisoning. For this reason also, mercuric chloride has been used to preserve wood. Its solution is decomposed by many metals, with the deposition of mercury. Mercurous chloride, Hg_2Cl_2, ordinarily called calomel, is precipitated whenever a chloride is added to a solution of a mercurous salt. It occurs native in tetragonal crystals. Commercially, it is prepared by subliming a mixture of mercuric sulphate, mercury, and salt:—

$$Hg(SO_4) + Hg + (NaCl)_2 = Na_2(SO_4) + Hg_2Cl_2.$$

The heavy white powder which condenses, is washed with water to remove any mercuric chloride. It is not soluble in water, though chlorine-water and nitric acid dissolve it by converting it into mercuric chloride. It volatilizes below a red heat, yielding four volumes of vapor; being decomposed into mercuric chloride and free mercury, each of which occupies two volumes; hence, the vapor of calomel turns gold-leaf white, and sublimed calomel always contains some corrosive sublimate. Calomel is gradually decomposed by light.

403. Mercuric Iodide, HgI_2, is precipitated as a brilliant scarlet, crystalline powder, on adding a solution of potassium iodide to one of a mercuric salt. The crystals are tetragonal octahedrons; but on heating to 200°, the iodide is volatilized and condenses in bright yellow orthorhombic plates, which on cooling, pass again into the red variety. This substance is therefore dimorphous, the two forms being probably polymeric. Mercurous iodide, Hg_2I_2, is obtained as a yellowish-green powder by precipitating a mercurous salt with potassium iodide.

404. Mercuric Oxide.—*Formula* HgO. When mercury is heated in the air to near its boiling point, red scales of mercuric oxide, commonly known as red precipitate, collect upon its surface. The same oxide is obtained by heating mercury nitrate until no more nitrous fumes are evolved. Mercuric oxide is soluble in acids, yielding mercuric salts. The di-meta-nitrate, $Hg(NO_3)_2$, aq., and the acid and the normal mono-meta-nitrates, $H_2Hg_2(NO_4)_2$, aq., and $Hg_3(NO_4)_2$, aq., are well-known compounds. The di-meta-sulphate, $Hg(SO_4)$, is decomposed by water, forming the yellow ortho-sulphate, $Hg_3(SO_6)$, turpeth mineral. Mercurous oxide, Hg_2O, is an unstable black powder, obtained by the action of alkali-hydrates upon calomel. With acids it forms mercurous salts.

405. Mercuric Sulphide, HgS, occurs native as the mineral cinnabar, both massive and in rhombohedral crystals. It is made artificially by direct synthesis. It is brilliant red in color, and is used as a pigment under the name of vermilion. Mercurous sulphide, Hg_2S, is a black powder known as ethiops mineral.

The mercury amides are very numerous. Mercuric chlor-amide, $Hg\left\{\begin{matrix}(NH_2)\\ Cl\end{matrix}\right.$, is known in pharmacy as white precipitate.

§ 3. Zinc.

Symbol Zn. *Atomic weight* 65. *Equivalence* II. *Density* 32·5. *Molecular weight* 65. *Molecular volume* 2. *One liter of zinc-vapor weighs* 2·91 *grams* (32·5 *criths*).

406. History and Occurrence.—An ore of zinc was used by the ancient Greeks in the manufacture of brass, under the name καδμία, a word said to be derived from Cadmus, who first taught them its use. The first zinc was made in Europe in the eighteenth century, it having been, until that time, imported from China. The name zinc was given by Paracelsus in the sixteenth century. Under the name calamine—a corruption of cadmia—two ores of zinc are now known; one a silicate, the calamine proper; the other a carbonate, now called smithsonite. Zinc-blende or sphalerite, and red zinc-ore, or zincite, are also well-known minerals.

407. Preparation and Properties.—Zinc is extracted from its ores by first roasting them in a reverberatory furnace, and then distilling them, mixed with charcoal, in close iron or earthen vessels. By the action of the charcoal, the oxygen is removed from the zinc oxide which constitutes the roasted ore, and the zinc thus set free, being volatile, distills over into suitable receivers. In Silesia, this distillation is effected in peculiarly-shaped muffles of fire-clay; in Belgium, in earthen-ware tubes; and in England, in fire-clay crucibles. The English zinc-furnace is represented in Fig. 99. It is conical in shape, and contains an interior dome, within which six crucibles are placed, each of which has a hole in the bottom, through which an iron pipe passes to convey away the zinc-vapor. The crucibles are charged with the mixture of roasted ore and coke, the covers are cemented on, and they are gradually raised to bright redness. At first the carbonous oxide burns at the mouth of the tube; but soon the greenish-white flame of zinc appears; the tube is then

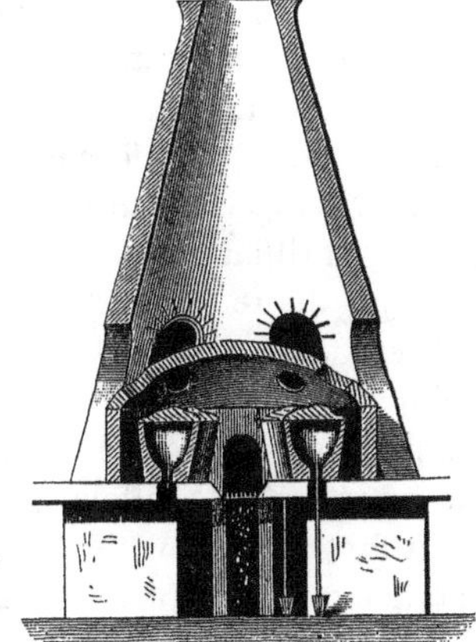

Fig. 99. English Zinc-furnace.

lengthened, and the condensed zinc is collected in suitable vessels placed beneath. It is afterward re-melted, cast into ingots, and sent into commerce under the name of spelter.

Zinc is a bluish-white, highly-crystalline metal, of specific gravity about 7·0. It is hard and brittle at ordinary temperatures, and also at 200°; but between 100 and 150°, it is malleable and ductile, and may be rolled into thin sheets. At 412° it melts, and at 1040°, it boils, evolving a vapor having half the normal density; its molecule is therefore mon-atomic. Zinc is oxidized readily by exposure to moist air; when strongly heated, it burns brilliantly, producing the oxide. It is acted upon by dilute acids, by the halogens, and by alkali-hydrates.

Zinc is used largely in the arts as sheet-zinc, for various purposes; it forms the positive element of all voltaic batteries. Its alloys are highly valuable; with copper it forms brass, with copper and tin, bronze and bell-metal, and with copper and nickel, german silver. Iron alloys readily with zinc on being plunged into a bath of the melted metal; the zinc-coating protects the iron by a galvanic action, whence iron thus treated is called galvanized iron.

COMPOUNDS OF ZINC.

408. Zinc Chloride, $ZnCl_2$, may be obtained either by the direct action of chlorine upon zinc, or by dissolving zinc in hydrochloric acid, and evaporating the solution. It is a white or grayish-white, translucent, crystalline substance, easily fusible, and volatile at a red heat. It is very deliquescent, and dissolves in water, forming a definite hydrate $ZnCl_2$, 2 aq., which crystallizes in octahedrons. It forms definite compounds with ammonia and with alkali-chlorides. It is used as a caustic in surgery; and in solution, as an antiseptic, and, with ammonium chloride, as a soldering-fluid. It is also a powerful de-hydrating agent.

409. Zinc Oxide, ZnO, occurs native as red zinc-ore or zincite. It is the sole product of the combustion of zinc in the air, and is made commercially by conducting zinc-vapor into chambers through which a current of air passes. It is a white powder, unalterable in the air, becoming transiently yellow on heating, and insoluble in water. It is used extensively as a pigment. Zinc hydrate, $Zn(OH)_2$, is obtained on adding an alkali-hydrate to a solution of a zinc-salt, as a white gelatinous mass, similar to alumina, but soluble in ammonia. It is strongly basic, forming well-defined salts with acids. Zinc sulphate, $ZnSO_4$, known as white vitriol, is obtained by oxidizing the sulphide, or by dissolving the metal in sulphuric acid. It crystallizes with seven molecules of water, in orthorhombic prisms. Zinc carbonate, $ZnCO_3$, occurs native as smithsonite; it is the principal ore of zinc. Zinc ortho-silicate, Zn''_2SiO_4, occurs native as willemite; and with a molecule of water Zn_2SiO_4, aq., is known as calamine.

410. Zinc Sulphide, ZnS, is found in nature in isometric crystals, forming the mineral sphalerite or blende. It is generally yellowish-brown in color, has a resinous luster, and is translucent. It is precipitated from zinc-solutions by alkali-sulphides, as a white precipitate, easily soluble in acids. The native sulphide is easily converted into the sulphate by roasting it with free access of air.

CADMIUM.—*Symbol* Cd. *Atomic weight* 112. *Equivalence* II. *Density* 56. *Molecular weight* 112. *Molecular volume* 2.

411. History and Preparation.—Cadmium was discovered in 1817 by Herrmann and also by Stromeyer; the latter gave it the name cadmium from cadmia, the ancient name of zinc-ore. Its sulphide is found native as greenockite. It occurs as an impurity in commercial zinc; but as it is more volatile than zinc, it comes over in the first products of its distillation; these are dissolved in acid, and the cadmium precipitated by zinc, in the metallic state.

412. Properties.—Cadmium resembles zinc, but is whiter, heavier, more easily fusible, and more volatile than that metal. Its specific gravity is 8·7, it is malleable and ductile at ordinary temperatures, but becomes brittle at 82°, and crackles when bent, like tin. It melts at 315°, and crystallizes in regular octahedrons on cooling. It is unalterable in the air, but at a red heat it burns, producing brown fumes of oxide. Acids act upon it slowly. Its compounds are analogous to those of zinc.

§ 4. MAGNESIUM.

Symbol Mg. *Atomic weight* 24. *Equivalence* II.

413. History and Occurrence.—Under the name of magnesia alba, magnesium carbonate was brought into Europe early in the eighteenth century; Black in 1755 showed that this substance contained a peculiar earth. The metal was prepared first by Davy in 1808; and in a purer form by Bussy in 1830. Bunsen and

Matthiessen in 1852, prepared it by electrolysis of the fused chloride. And in 1857, Deville and Caron obtained it in large quantity by acting on the chloride with sodium; a process subsequently improved by Sonstadt. Magnesium occurs abundantly in nature, but always in combination. Its hydrate forms the mineral brucite; its carbonate, magnesite; its sulphate, epsomite; its borate, boracite; its fluo-phosphate, wagnerite, etc. Numerous magnesian silicates are also known; and the double carbonate of magnesium and calcium, or dolomite, exists abundantly as magnesian limestone. Magnesium chloride exists in many natural waters, especially in sea-water.

414. Preparation and Properties.—Magnesium is prepared on the large scale by heating to bright redness a mixture of six parts of magnesium chloride, one of sodium chloride, one of calcium fluoride, and one of sodium, in a covered crucible. The globules of magnesium thus obtained, are purified by distillation in an atmosphere of hydrogen.

Magnesium is a silver-white, brilliant metal, having a specific gravity of 1·75. It is somewhat malleable and ductile, but is not very tenacious. It melts at a red heat, and may be obtained crystallized in regular octahedrons. At higher temperatures, it is volatile. It is permanent in dry air, though it oxidizes readily in moist air. Heated to redness it takes fire and burns with a dazzling light, very rich in chemical rays. A wire 0·297 m.m. in thickness gives a light equal to that of 74 stearin candles, five of which weigh a pound. Acids attack it with ease, and it unites directly with most negative elements, including nitrogen. It has been used of late years as a source of light in photography.

COMPOUNDS OF MAGNESIUM.

415. Magnesium Chloride, $MgCl_2$, is best prepared by igniting the double chloride of magnesium and ammonium; the ammonium chloride is thereby driven off, and the magnesium chloride is left as a white, translucent, crystalline mass, readily fusible, and somewhat volatile. It is deliquescent, and dissolves in water, forming the hydrate $MgCl_2$, 6 aq. The mineral carnallite is a potassio-magnesium chloride, $KMgCl_3$, 6 aq.

416. Magnesium Oxide, MgO, commonly called magnesia, occurs native as periclasite, crystallized in isometric forms. It is the sole product of the combustion of magnesium in air, and is left whenever its carbonate, hydrate, or nitrate is ignited. It is an infusible, insoluble, bulky, white powder, which gradually unites with water to form the hydrate. Magnesium hydrate, $Mg(OH)_2$, is precipitated from solutions of magnesium salts by alkali-hydrates. It is slightly soluble in water, and reacts alkaline to test-papers. It occurs in nature as brucite, crystallized in rhombohedrons. Magnesium sulphate, $MgSO_4$, may be prepared by dissolving the oxide, hydrate, or carbonate in sulphuric acid. It is obtained commercially, either from sea-water, from dolomite,—a calcio-magnesian carbonate,—or from serpentine, which is a magnesium silicate. It occurs native as epsomite, $MgSO_4$, 7 aq., and is found also in the waters of bitter saline springs, as those of Epsom in England; whence the name Epsom salt, applied to this substance. It crystallizes in orthorhombic prisms, containing seven molecules of crystal-water. They dissolve in three times their weight of cold water. Magnesium carbonate, $MgCO_3$, is found in nature as the mineral magnesite. Magnesia alba, the common commercial form of this substance, is obtained by precipitating the sulphate with sodium carbonate. It is a light, white powder, soluble in acids, even the carbonic, and having the composition $H_4Mg_4(CO_4)_3$, 2 aq. It occurs native as hydromagnesite.

§ 5. CALCIUM.

Symbol Ca. *Atomic weight* 40. *Equivalence* II *and* IV.

417. History and Occurrence.—Calcium carbonate and sulphate were known to the ancients, the former being burned into lime for making mortar. It is from the Latin name for lime, calx, that the word calcium is derived. The metal itself was obtained in an impure form by Davy in 1808; and in 1855, Matthiessen prepared it pure and in considerable quantity. Calcium is one of the most abundant of the elements. As carbonate, it forms the mineral calcite, and the rock-masses known as limestone, chalk, and marble. As sulphate, it forms vast beds of gypsum; as phosphate it occurs as apatite; as fluoride, in fluor spar; and as silicate, it is found in numerous minerals. It forms an important constituent of the animal skeleton, whether external or internal, and is essential to vegetable growth.

418. Preparation and Properties.—Calcium may be prepared either by the electrolysis of its chloride, previously mixed with two-thirds its weight of strontium chloride, and fused in a porcelain crucible; or by igniting the iodide with sodium in a closed vessel. It is a light-yellow, brilliant metal, of specific gravity 1·57, very malleable and ductile, and about as hard as gold. It is permanent in dry air, but is oxidized if the air be moist. At a red heat it burns vividly.

COMPOUNDS OF CALCIUM.

419. Calcium Chloride, $CaCl_2$, is produced by dissolving the oxide or carbonate in hydrochloric acid, and in many other ways; it is therefore frequently a waste product in the chemical arts. On evaporating its solution in water, hexagonal crystals of the hydrate, $CaCl_2$, 6 aq., are obtained, which lose two molecules of water when dried in vacuo, and the whole at 200°. The crystals are used with snow as a freezing mixture; in it a thermometer falls to —48·5°. Calcium chloride melts at a full red heat; it has a strong attraction for water, and is deliquescent; it is hence used for drying gases.

420. Calcium Oxide, CaO, commonly known as lime, is always prepared on the large scale by igniting its carbonate. This operation is conducted in a rather rude furnace of masonry called a kiln, Fig. 100, built usually upon the side of a hill. An arch of the limestone is first thrown across just above the fire, and upon this the charge rests. The fuel employed is generally wood, and the fire is maintained for three or four days, until the whole mass is converted into lime. The purer the limestone, the better the product; perfectly pure lime is obtained by igniting crystallized calcite.

Calcium oxide is a white, hard, infusible substance, of specific gravity 2·3 to 3. When moistened with water, it unites directly with it, with the evolution of great heat, to form the hydrate. This process is called slaking. Calcium hydrate, $Ca(OH)_2$, is a soft, white, bulky powder, soluble in 530 times its weight of cold water, forming an alkaline, feebly-caustic liquid, known as lime-water, which yields on evaporation in vacuo, hexagonal prisms of the hydrate. It absorbs carbonic di-oxide readily. Calcium sulphate, $CaSO_4$, occurs native as anhydrite; and crystallized with two molecules of water, as selenite. It is also found massive

Fig. 100. Lime-kiln.

as gypsum. It is precipitated on adding a sulphate to a concentrated solution of a calcium salt. By heating gypsum to 250°, it parts with its water, producing what is called plaster of Paris. The burned gypsum on being mixed with water, again unites with it and "sets" or becomes hard. Hence its use as a cement. Calcium carbonate, $CaCO_3$, is an abundant mineral. It is dimorphous, being orthorhombic in aragonite, and rhombohedral in calcite. It occurs also more or less pure, as marble, chalk, and limestone. In water containing carbonic acid, it is quite soluble; such waters are called hard. Calcium phosphate, $Ca_3(PO_4)_2$, is precipitated from a solution of a calcium salt by sodium phosphate in excess. It occurs as a fluo-phosphate in the mineral apatite, and forms the chief constituent of the bones of animals. Two acid calcium phosphates, $HCaPO_4$, and $H_4Ca(PO_4)_2$, are known, the former with 2 aq., constituting the mineral brushite.

§ 6. STRONTIUM AND BARIUM.

STRONTIUM.—*Symbol* Sr. *Atomic weight* 87·5. *Equivalence* II *and* IV.

421. History and Occurrence.—Strontium was distinguished as a peculiar substance by Hope in 1792. The metal was first prepared pure by Bunsen and Matthiessen in 1855. It occurs in nature both as sulphate or celestite, and as carbonate or strontianite. From the latter name, which is taken from that of Strontian, in Scotland, where the mineral was first observed, the name strontium comes.

422. Preparation and Properties. — Metallic strontium is prepared by the electrolysis of its chloride. It is a pale-yellow metal, of specific gravity 2·54, harder than lead, melting at a red heat, and burning vividly if exposed to the air. It is quite permanent in dry air, but decomposes water readily, evolving hydrogen.

COMPOUNDS OF STRONTIUM.

423. Strontium Chloride, $SrCl_2$, is obtained as a hydrate $SrCl_2$, 3 aq., by dissolving the oxide or carbonate in hydrochloric acid. On evaporation, deliquescent crystals are obtained, which lose their water on fusion, leaving a colorless, glassy mass, soluble in water and in alcohol.

424. Strontium Oxide, SrO, is generally prepared by igniting the nitrate. It is a grayish-white, porous mass, is infusible, and unites energetically with water to form the hydrate. Strontium hydrate, $Sr(OH)_2$, prepared as above, is soluble in water, forming an alkaline liquid which, on evaporation, deposits crystals of $Sr(OH)_2$, 8 aq. Strontium nitrate, $Sr(NO_3)_2$, is employed in pyrotechny for producing a crimson fire. Strontium di-oxide, $Sr^{IV}O_2$, is precipitated as a hydrate by adding hydrogen peroxide to a solution of strontium hydrate.

BARIUM.—*Symbol* Ba. *Atomic weight* 137. *Equivalence* II *and* IV.

425. History and Occurrence.—Barium was first recognized as a new element by Scheele in 1774. Davy in 1808, first isolated the metal. The compounds of barium which occur native are the sulphate, or barite, and the carbonate, or witherite. The former from its weight was formerly called heavy-spar; hence the name barium, from βαρύς, heavy.

426. Preparation and Properties.—Barium, obtained by the electrolysis of its chloride, is a yellow, lustrous, malleable metal, of specific gravity 4·0, which decomposes water rapidly.

429. Compounds of Barium.—Barium chloride, $BaCl_2$, is usually obtained by dissolving the carbonate, or the sulphide—prepared from the native sulphate—in hydrochloric acid, and crystallizing. Orthorhombic crystals separate, having the formula $BaCl_2$, 2 aq., which lose all their water at 100°. Barium oxide, BaO, is a grayish-white, infusible mass, obtained by heating the nitrate to redness. With water it slakes, forming a hydrate, $Ba(OH)_2$, which crystallizes from a hot saturated solution with 8 aq. Barium nitrate, $Ba(NO_3)_2$, is used in the green fire of pyrotechny, for which, however, the chlorate, $Ba(ClO_3)_2$, is to be preferred. Barium sulphate, $BaSO_4$, occurs native as barite; when heated with charcoal, it is reduced to barium sulphide, BaS. Barium di-oxide, BaO_2, is produced by heating the oxide in a stream of oxygen. It is used in the preparation of hydrogen peroxide.

EXERCISES.

1. What weight of copper may be obtained from 1,000 kilograms of melaconite? Of chalcocite? Of malachite?
2. Calculate the composition of crystallized cupric sulphate.
3. One c.c. mercury will give what volume of vapor?
4. How much mercuric sulphate, and how much salt, are required to give one kilogram of $HgCl_2$? Of Hg_2Cl_2?
5. Calculate the percentage composition of HgO. Of Hg_2O.
6. Smithsonite, containing 75 per cent of zinc carbonate, yields what weight of zinc to the kilogram?
7. 500 kilograms of blende yield when roasted, what weight of crystallized zinc sulphate?
8. To yield 5 c.c. of magnesium, requires how much magnesite?
9. What is the relative length of two bars 5 centimeters in diameter, one of platinum, the other of magnesium, each of which weighs a kilogram?
10. What weight does limestone lose in burning?
11. Compare the percentage composition of anhydrite and selenite.
12. Which contains most oxygen, strontianite or witherite?

CHAPTER ELEVENTH.

POSITIVE MONADS.

§ I. SILVER.

Symbol Ag. *Atomic weight* 108. *Equivalence* I *and* III.

428. History and Occurrence.—Silver has been known in the metallic form from the earliest historic times. It occurs native, both crystallized and massive, and is found also in combination, as sulphide, in argentite ; as sulph-antimonite, in pyrargyrite, miargyrite, and stephanite ; as chloride in cerargyrite ; as bromide in bromyrite ; as chloro-bromide in embolite, etc.

429. Preparation.—The process employed for the extraction of silver varies with the quality of the ore. At Freiberg, the ore—an impure sulphide—is roasted with ten per cent of salt, the resulting mass is ground to a fine powder, and agitated in revolving barrels containing water and scrap iron, by which the silver chloride is reduced to the metallic state. Mercury is then added to dissolve the silver, and by distilling the amalgam thus obtained, the silver is left pure. A crude modification of this process is employed in Mexico and Chili. Much of the lead of commerce is argentiferous. To extract the silver from it, the lead is melted in large iron pots, and allowed to cool gradually. Crystals of pure lead separate, while the alloy of silver and lead remains fluid ; these crystals are removed and re-melted, and again allowed to cool; so that ultimately, by a continuation of the process, a very rich alloy is obtained which may contain 300 ounces of silver to the ton. This alloy is then cupelled; i. e., it is melted in a reverberatory furnace, whose hearth is composed of bone-ash, and the lead oxidized by allowing a current of air to pass over it; the lead oxide thus formed fuses and is absorbed by the bone-ash, while the silver, being unaltered, is finally left pure.

430. Properties.—Silver is a remarkably white, brilliant metal, of specific gravity 10·5. It is harder than gold, but may be hammered into leaves only one four-thousandth of a millimeter thick, and drawn into wire so fine that two thousand meters would weigh but one gram. It has a high tenacity, the weight of 85 kilograms being required to break a wire of silver two millimeters in diameter. It is the best conductor of heat and electricity known. It is fusible at about 1000°, and is slightly volatile. It may be obtained in isometric crystals by slow cooling. When melted, it is capable of absorbing twenty-two times its volume of oxygen gas, which is again evolved when it solidifies. It is unaltered in the air at any temperature, though it is readily acted on by chlorine, by sulphur, and by phosphorus. Nitric acid dissolves it easily, sulphuric and hydrochloric acids with difficulty. It is not attacked by melted niter, nor by fused alkali-hydrates.

431. Uses.—Owing to its softness, silver is rarely used alone. It is generally alloyed with copper, which, while it increases its hardness, scarcely injures its color. The coin-alloy of the United States and France contains ten per cent of copper; that of England, 7·5, and that of Germany 12·5 per cent. The silver used in silver-plate, contains usually from 70 to 95 per cent of pure silver.

COMPOUNDS OF SILVER.

432. Silver Chloride, AgCl, occurs native as cerargyrite. It may be obtained by the direct union of silver and chlorine, or by precipitating a solution of silver nitrate by a chloride. A white curdy mass, soluble in ammonium hydrate, but insoluble in nitric acid, is thrown down, which on drying becomes a white powder. When heated it fuses, and on cooling, solidifies to a crystalline, translucent, sectile mass resembling horn, whence the name horn-silver, sometimes applied to it. It has a specific gravity of 5·4, crystallizes in isometric forms, and turns black on exposure to light. For this latter reason it is used in photography.

433. Silver Oxide, Ag_2O, is usually prepared by adding a strong, hot solution of silver nitrate to one of potassium hydrate. It is also the product of the combustion of silver at high temperatures. It is a dark-brown or black powder, of specific gravity 7·2, easily decomposed by heat, and partially by light. It is scarcely soluble in water, though ammonia dissolves it readily, the solution depositing a violently-explosive crystalline compound, probably the nitride, Ag_3N, on exposure to the air. Silver hydrate, Ag(OH), is a strong base, reacts alkaline, and becomes silver oxide on heating to 60°. Silver nitrate, $AgNO_3$, is prepared by dissolving silver in nitric acid, and evaporating to crystallization. Transparent orthorhombic prisms are obtained, which are soluble in their own weight of cold water. They fuse at a moderate heat, and at a higher temperature are decomposed. Silver nitrate blackens in presence of organic matter; it is used therefore as a hair-dye, and for indelible ink. Cast in sticks, it is employed as a caustic in surgery. Other salts of silver are the sulphate, Ag_2SO_4, and the phosphate, Ag_3PO_4. Silver di-oxide, or per-oxide, Ag_2O_2, is obtained by electrolysis.

§ 2. Lithium.

Symbol Li. *Atomic weight* 7. *Equivalence* I.

434. History and Occurrence.—Lithium oxide was recognized first as a peculiar substance by Arfvedson in 1817. The metal was first prepared pure by Bunsen and Matthiessen in 1855. Lithium is a rare substance, being found principally in the minerals amblygonite, spodumene, petalite, lepidolite, and triphylite. The water of a mineral spring in Cornwall contains the chloride in considerable quantity. Traces of lithium have been detected by the spectroscope in sea-water, in many minerals, in tobacco-ashes, and even in meteorites.

435. Preparation and Properties.—Lithium is best obtained by the electrolysis of the fused chloride. It is a brilliant, silver-white metal, somewhat softer than lead, and remarkably light, having a specific gravity of only 0·578. It melts at 180° and burns in the air with an intense white flame when more strongly heated. It tarnishes in the air, and decomposes water, evolving hydrogen. Its compounds are the chloride, LiCl, a deliquescent, fusible, and volatile salt; the oxide, Li_2O, and hydrate, Li(OH), the latter a caustic, strongly alkaline substance; the carbonate, Li_2CO_3, the sulphate, Li_2SO_4, and the phosphate, Li_3PO_4, all of which are well-defined and characteristic salts.

§ 3. Ammonium.

Symbol Am. *Formula* (NH_4). *Molecular weight* 18. *Equivalence* I.

436. History.—The salts which ammonia forms by direct union with acids, have a remarkable similarity to those formed by the metals of this group. To account for this resemblance, Berzelius, in 1816, acting upon a theory of Ampère, proposed to consider these salts as compounds of ammonium, a monad compound radical, $(N^VH_4)'$, capable of acting like the monad elements sodium and potassium. Being unsaturated, ammonium, if it exist free must exist as a double molecule, N_2H_8. Weyl, by condensing ammonia gas in presence of sodium, obtained a bright blue metallic-like liquid, which he assumed to be sod-ammonium, $N_2H_6Na_2$. By acting upon ammonium chloride with this, a similar blue liquid was obtained, which he considered to be free ammonium. Moreover, when mercury containing one per cent of sodium is placed in a saturated solution of ammonium chloride, it swells prodigiously in bulk, becoming a pasty mass like butter. This is the so-called ammonium amalgam. It rapidly decomposes, yielding ammonia and hydrogen gases.

437. Compounds of Ammonium.—Ammonium chloride, $(NH_4)Cl$, found native as sal-ammoniac, is prepared on the large scale from the ammoniacal liquors of the gas-works, by adding hydrochloric acid, crystallizing out the crude chloride, and purifying by sublimation. It is thus obtained in white fibrous masses, soluble in 2·72 parts of water at 18°, and crystallizing in cubes and regular octahedrons. Ammonium sulphate, $(NH_4)_2SO_4$, is also obtained from gas-liquor.

It occurs in transparent, orthorhombic crystals, isomorphous with potassium sulphate. Ammonium nitrate, $(NH_4)NO_3$, is prepared by neutralizing nitric acid with ammonium hydrate, and crystallizing. It is used for making hyponitrous oxide gas. Commercial ammonium carbonate is a mixture of acid carbonate and carbamate, $(H(NH_4)CO_3)_2$ and $(NH_4)(NH_2CO_2)$. On exposure to air, the carbamate is volatilized.

§ 4. Sodium.

Symbol Na. *Atomic weight* 23. *Equivalence* I *and* III.

438. History and Occurrence.—Sodium oxide was first clearly distinguished from potassium oxide by Duhamel in 1736. The metal was first obtained by Davy in 1807. Sodium is one of the most abundant of the elements. Its chloride, or salt, known as the mineral halite, is found not only in immense deposits of rock-salt, but also in enormous quantities in sea-water, and the water of saline springs. Sodium also occurs in the form of nitrate, or soda-niter; of borate, or borax; of carbonate, or trona; and of silicate, in albite, oligoclase, sodalite, etc. It is found in marine plants and is essential to animal life.

439. Preparation and Properties.—Sodium is prepared by reducing its oxide by carbon at a white heat, thus:—

$$Na_2O + C = Na_2 + CO.$$

Practically, thirty kilograms of dry sodium carbonate, thirteen kilograms of charcoal, and three kilograms of chalk, are intimately mixed together, calcined, and introduced into iron cylinders heated in a reverberatory furnace, Fig. 101. At a bright red heat, the sodium distills over, and is collected in the flat receiver shown in the figure. To purify it, it is re-distilled, melted under petroleum, and cast into ingots, which are preserved under naphtha.

Sodium is a lustrous, silver-white, soft metal, of specific gravity 0·98, becoming brittle at −20°, and fusing at

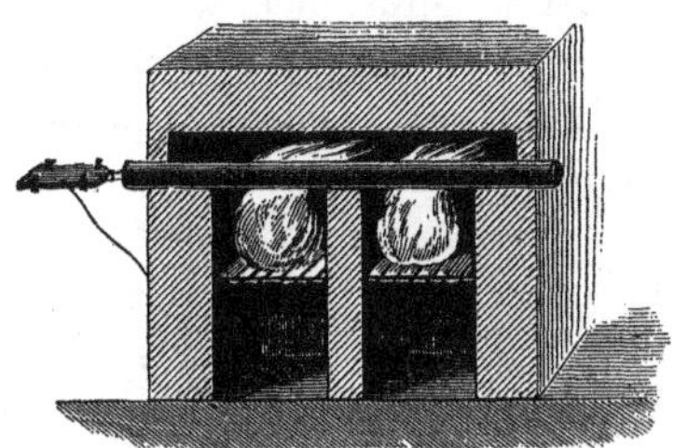

Fig. 101. Sodium-furnace.

97°. It crystallizes in tetragonal octahedrons. On exposure to air it rapidly tarnishes, and if thrown on water, decomposes it with effervescence; if it be prevented from moving, or if the water be warm, it takes fire, burning with a characteristic yellow flame. It is used in metallurgy as a reducing agent.

COMPOUNDS OF SODIUM.

440. Sodium Chloride, NaCl, may be formed by the direct union of its constituents, as by burning sodium in chlorine gas. It is obtained commercially, either by mining it directly, in which form it is known as rock salt, or by evaporating the water of saline springs, producing boiled salt if artificial heat be used; or solar salt if the heat be natural. Large quantities of salt are obtained from sea-water, of which it constitutes 2·75 per cent. Sodium chloride is a colorless, transparent solid, crystallizing in cubes, and having a specific gravity of 2·15,

It has an agreeable saline taste, is deliquescent in moist air, and is soluble in three times its weight of water.

441. Sodium Oxide, Na_2O, **and Hydrate,** $Na(OH)$. Sodium oxide may be obtained by the combustion of the metal in air, or by heating the hydrate with sodium. It is a white, fusible substance which unites directly with water to form the hydrate. Sodium hydrate, known commonly as caustic soda, is made in the pure form by the action of sodium upon water. Commercially, it is prepared by the action of calcium hydrate —milk of lime—upon sodium carbonate; the clear liquid thus obtained is evaporated in vessels of iron or of silver, and the fused mass which is left is poured on to flat plates, or cast into sticks. It is a white, opaque, brittle solid, of specific gravity 2·00. It is deliquescent and absorbs carbonic di-oxide from the air, forming the carbonate. It may be obtained crystallized in monoclinic prisms, having the composition $(NaOH)_2$, 7 aq. Sodium peroxide, Na_2O_2, is produced by heating sodium in oxygen, as a white, friable, deliquescent mass.

442. Sodium Sulphate, Na_2SO_4, is obtained abundantly commercially, as a residue in many chemical processes, as in the preparation of nitric and hydrochloric acids. It is also largely produced as an intermediate product in the soda manufacture. It occurs native, anhydrous as thenardite, and hydrated as mirabilite. It crystallizes from solution in large colorless monoclinic prisms, which have the composition Na_2SO_4, 10 aq., and which are efflorescent in dry air, losing all their water. The anhydrous salt is soluble in two and a half parts of water at 100°. The acid salt, hydro-sodium sulphate, $HNaSO_4$, is formed by adding sulphuric acid to the normal sulphate.

443. Sodium Carbonate, Na_2CO_3, was formerly extracted from the ashes of marine plants. It is now produced in immense quantities from salt, by a process proposed by Leblanc. This process consists, 1st, in treating the salt with sulphuric acid, by which sodium sulphate is produced; and 2d, in heating the sodium sulphate with coal and limestone, in a reverberatory furnace, by which sodium carbonate and calcium sulphide are obtained in the form of a black mass called black-ash. This is extracted with water, evaporated to dryness, calcined with sawdust, and brought into commerce as soda-ash. By solution in water and evaporation, sodium carbonate is obtained crystallized, Na_2CO_3, 10 aq. The crystals are efflorescent, are readily soluble in water, have an alkaline reaction, and a nauseous taste. The acid salt, hydro-sodium carbonate, $HNaCO_3$, is prepared by exposing crystals of the normal salt to carbonic di-oxide gas.

444. Sodium Nitrate, $NaNO_3$, is brought from Peru, where it occurs native, under the name soda saltpeter. It crystallizes in rhombohedrons and is deliquescent. Normal sodium phosphate, Na_3PO_4, is unstable. Hydro-di-sodium phosphate, HNa_2PO_4, is alkaline in its reaction, while di-hydro-sodium phosphate, H_2NaPO_4 reacts acid.

§ 5. Potassium.

Symbol K. *Atomic weight* 39. *Equivalence* I, III, *and* V.

445. History and Occurrence.—The lye obtained by leaching the ashes of land-plants has long been used in soap-making; and the solid product of the evaporation of this lye has long been known in commerce as potashes, from its origin. The metal potassium was first obtained by Davy in 1807, by the aid of electricity; but it was soon after prepared chemically by Gay-Lussac and Thenard. Potassium occurs somewhat abundantly in nature, but always in combination. In the mineral kingdom, it is found as nitrate or niter, as chloride or sylvite, as potassio-magnesium chloride, or carnallite, and as sulphate, or aphthitalite. It exists also in orthoclase and in muscovite, and in the waters of the ocean and mineral springs. Land-plants contain it largely, and it is essential to animal life.

446. Preparation and Properties. — Potassium is now prepared by calcining an intimate mixture of the carbonate with charcoal, obtained by igniting the tartrate. The process is quite similar to that described for sodium; but owing to the tendency of the metal to unite directly with the carbonous oxide to produce a dangerously-explosive body, it is far more difficult. Potassium is a soft, brilliant, bluish-white metal, of specific gravity 0·865, becoming brittle at 0°, and at 62·5° melting to a liquid resembling mercury. From this it may be crystallized in tetragonal octahedrons. It tarnishes instantly in the air, and must therefore be preserved under naphtha. Thrown on water, it at once decomposes it, evolving so much heat that the hydrogen set free takes fire and burns with a characteristic violet flame, Fig. 102. It unites actively with chlorine and with sulphur.

Fig. 102. Potassium on Water.

COMPOUNDS OF POTASSIUM.

447. Potassium Chloride, KCl, constitutes the mineral sylvite. It is obtained commercially from sea-water, or from an abundant mineral of the Stassfurt mines, carnallite. It is a transparent colorless solid, which crystallizes in cubes, has a specific gravity of 1·9, and tastes like common salt. It decrepitates when heated, melts at a red heat, and is volatile at higher temperatures. It dissolves in about three times its weight of water at 15°, producing great cold.

Potassium iodide, KI, is prepared on the large scale by the direct action of iodine upon potassium hydrate :—

$$\underset{\textit{Iodine.}}{(I_2)_3} + \underset{\textit{Potassium hydrate.}}{(KOH)_6} = \underset{\textit{Potassium iodide.}}{(KI)_5} + \underset{\textit{Potassium iodate.}}{KIO_3} + \underset{\textit{Water.}}{(H_2O)_3}.$$

By evaporation of the solution, and gentle ignition of the residue, the iodate is converted into iodide. Potassium iodide occurs in cubical crystals, which are deliquescent in moist air. It is readily soluble in water and alcohol, and has a specific gravity of 3. Potassium bromide, KBr, is quite similar in properties to the iodide, and is obtained by an analogous process.

448 Potassium Oxide, K_2O, **and Hydrate,** K(OH).— Potassium oxide is obtained by the direct oxidation of potassium or by the action of potassium upon the hydrate. It is a white, deliquescent, and caustic substance, which unites energetically with water to form the hydrate. This hydrate, commonly called caustic potash, is generally prepared by the action of milk of lime — calcium hydrate — upon potassium carbonate. The solution may be evaporated, and the fused hydrate thus left be cast into sticks, the form in which it is usually found in commerce. It is a white, opaque, deliquescent, brittle mass, of specific gravity 2·1, and freely soluble in water and alcohol. Its solution is powerfully alkaline, turning reddened litmus back again to blue, neutralizing completely the strongest acids, and acting readily upon the skin. In the solid form, it is sometimes used as a cautery. Potassium also forms a di-oxide, K_2O_2, and a tetr-oxide, K_2O_4.

449. Potassium Salts. — Potassium chlorate, $KClO_3$, is prepared by passing chlorine through a solution of potassium chloride containing milk of lime :—

$$KCl + (Ca(OH)_2)_3 + (Cl_2)_3 = (CaCl_2)_3 + (H_2O)_3 + KClO_3.$$

By evaporation of the solution, the chlorate crystallizes out. Potassium sulphate, K_2SO_4, is obtained as a residue in the preparation of nitric acid from niter, and is a product also of the evaporation of sea-water. It forms hard orthorhombic crystals, which have a bitter taste, and dissolve in ten parts of cold water. Potassium nitrate, KNO_3, occurs as an efflorescence on the surface of the soil in Bengal. It is extracted by solution in water, and evaporation, and is brought into commerce as crude saltpeter. It is also obtained by decomposing native sodium nitrate, or native or artificial calcium nitrate by potassium carbonate. It crystallizes generally in transparent orthorhombic prisms, which are not deliquescent, have a cooling taste, and dissolve readily in water. It fuses below redness to a colorless liquid, and at a high temperature is decomposed, evolving oxygen. It deflagrates with combustibles, and is used in making gunpowder. Potassium carbonate, K_2CO_3, is the essential constituent of the crude potash of commerce, which is obtained by evaporating the leachings of wood-ashes. When refined it is known as pearlash. It forms a white, granular, very deliquescent mass, having an alkaline reaction, and fusible at full redness. It is soluble in less than its own weight of water, but is insoluble in alcohol. The acid salt, hydro-potassium carbonate, $HKCO_3$, is permanent in the air, and has a faintly alkaline reaction.

§ 6. Rubidium and Cæsium.

Rubidium.—*Symbol* Rb. *Atomic weight* 85. *Equivalence* I.

450. Preparation and Properties.—Rubidium was first detected in the water of the Dürckheim mineral spring, by Bunsen in 1860, by means of the spectroscope. Its spectrum contains two characteristic dark red lines; hence its name, from the Latin rubidus, dark-red. By distilling the carbonate with charcoal, rubidium is obtained as a soft white metal, of specific gravity 1·5. It melts at 101°, and volatilizes below a red heat. It is more easily oxidized than potassium, and takes fire in the air, burning with a violet flame.

451. Rubidium Salts.—The salts of rubidium resemble very closely those of potassium. The chloride crystallizes in cubes, dissolves in its own weight of water at 150°, and forms a double salt with platinum chloride. The nitrate, $RbNO_3$, resembles saltpeter, but crystallizes in hexagonal prisms. The carbonate, Rb_2CO_3, is an alkaline deliquescent salt. The sulphate, Rb_2SO_4, is isomorphous with potassium sulphate.

CÆSIUM.—*Symbol* Cs. *Atomic weight* 133. *Equivalence* I.

452. History.—Cæsium was discovered at the same time with rubidium, and in the same mineral water. The name cæsium comes from cæsius, sky-blue, and has reference to two bright blue lines in its spectrum. The Dürckheim water contains in five kilograms scarcely one milligram of cæsium chloride; but the lepidolite of Hebron in Maine contains 0·3 per cent of cæsium, and the rare mineral pollucite contains 32 per cent. Metallic cæsium has not yet been obtained. The salts of cæsium resemble those of rubidium.

EXERCISES.

1. 100 parts of silver form by union with chlorine, 132·84 parts of silver chloride; what is the atomic weight of silver?
2. How is silver extracted from argentiferous lead?
3. How much silver is needed to make 250 grams silver nitrate?
4. Give the theory of ammonium. Illustrate it.
5. What volume of ammonia gas in one kilogram of $(NH_4)_2SO_4$?
6. Twenty kilograms Na_2CO_3 yield how many c.c. of sodium?
7. One c.c. rock-salt contains what weight of sodium? What volume of chlorine?
8. Write the reactions in the soda-process of Leblanc.
9. 100 kilograms of salt yield what weight of sodium carbonate?
10. By decomposing a kilogram of $NaNO_3$ by K_2CO_3 what weight of KNO_3 is obtained?
11. How much water will one gram of potassium decompose? One gram of sodium?
12. By whom were rubidium and cæsium discovered? When?

Appendix and Index

APPENDIX.

TABLE I. — THE METRIC SYSTEM.

Measures of Length.

Millimeter	0.001	**of a meter.**	**0·0394**	**inch.**
Centimeter	**0·01**	" "	**0·3937**	"
Decimeter	**0·1**	" "	**3.9370**	**inches.**
METER	**1**	**meter.**	**39.37**	"
Dekameter	10	meters.	393·7	"
Hectometer	100	"	328	feet, 1 inch.
Kilometer	1000	"	3280	" 10 inches.
Myriameter	10000	"	6·2137	miles.

Measures of Surface.

Centare	1	square meter.	1550	square inches.
ARE	100	square meters.	119·6	square yards.
Hectare	10000	square meters.	2·471	acres.

Measures of Capacity.

Milliliter	**0·001**	**of a liter.**	**0·27**	fluid drachm.
Centiliter	0·01	" "	0·338	fluid ounce.
Deciliter	0·1	" "	0·845	gill.
LITER	1	cubic decimeter.	1·0567	quarts.
Dekaliter	10	cubic decimeters.	2·6417	gallons.
Hectoliter	100	" "	26·417	"
Kiloliter (stére)	1000	" "	264·17	"

Weights.

Milligram	0·001	wt.	1 cu. m.m.	water.	0·0154 grain Av.
Centigram	0·01	"	10 " "	"	0·1543 " "
Decigram	0·1	"	0.1 " c.m.	"	1·5432 grains "
GRAM	1	"	1 c.c.	"	15·432 " "
Dekagram	10	"	10 "	"	0·3527 ounce "
Hectogram	100	"	1 deciliter	"	3·5274 ounces "
Kilogram	1000	"	1 liter	"	2·2046 pound "
Myriagram	10000	"	10 liters	"	22·046 " "
Quintal	100000	"	1 hectoliter	"	220·46 " "
Tonneau	1000000	"	1 cu. meter	"	2204·6 " "

COMPARISON OF CENTIGRADE AND FAHRENHEIT DEGREES.

Cent.	*Fahr.*	*Cent.*	*Fahr.*	*Cent.*	*Fahr.*	*Cent.*	*Fahr.*
—40	—40·0	—5	+23·0	+30	+86·0	+65	+149·0
39	38·2	4	24·8	31	87·8	66	150·8
38	36·4	3	26·6	32	89·6	67	152·6
37	34·6	2	28·4	33	91·4	68	154·4
36	32·8	—1	30·2	34	93·2	69	156·2
35	31·0	0	32·0	35	95·0	70	158·0
34	29·2	+1	33·8	36	96·8	71	159·8
33	27·4	2	35·6	37	98·6	72	161·6
32	25·6	3	37·4	38	100·4	73	163·4
31	23·8	4	39·2	39	102·2	74	165·2
30	22·0	5	31·0	40	104·0	75	167·0
29	20·2	6	42·8	41	105·8	76	168·8
28	18·4	7	44·6	42	107·6	77	170·6
27	16·6	8	46·4	43	109·4	78	172·4
26	14·8	9	48·2	44	111·2	79	174·2
25	13·0	10	50·0	45	113·0	80	176·0
24	11·2	11	51·8	46	114·8	81	177·8
23	9·4	12	53·6	47	116·6	82	179·6
22	7·6	13	55·4	48	118·4	83	181·4
21	5·8	14	57·2	49	120·2	84	183·2
20	4·0	15	59·0	50	122·0	85	185·0
19	2·2	16	60·8	51	123·8	86	186·8
18	—0·4	17	62·6	52	125·6	87	188·6
17	+1·4	18	64·4	53	127·4	88	190·4
16	3·2	19	66·2	54	129·2	89	192·2
15	5·0	20	68·0	55	131·0	90	194·0
14	6·8	21	69·8	56	132·8	91	195·8
13	8·6	22	71·6	57	134·6	92	197·6
12	10·4	23	73·4	58	136·4	93	199·4
11	12·2	24	75·2	59	138·2	94	201·2
10	14·0	25	77·0	60	140·0	95	203·0
9	15·8	26	78·8	61	141·8	96	204·8
8	17·6	27	80·6	62	143·6	97	206·6
7	19·4	28	82·4	63	145·4	98	208·4
—6	+21·2	+29	+84·2	+65	+147·2	99	210·2
						+100	+212·0
+110	+230	+210	+410	+310	+590	+410	770
120	248	220	428	320	608	420	788
130	266	230	446	330	626	430	806
140	284	240	464	340	644	440	824
150	302	250	482	350	662	450	842
160	320	260	500	360	680	460	860
170	338	270	518	370	698	470	878
180	356	280	536	380	716	480	896
190	374	290	554	390	734	490	914
+290	+392	+300	+572	+400	752	+500	+932
+500	+932	+800	+1472	+1100	+2012	+1400	+2552
600	1112	900	1652	1200	2192	1500	2732
+700	+1292	+1000	+1832	+1300	+2372	+1600	+2912

TABLE III.—SOLUBILITIES OF CHEMICAL COMPOUNDS.

	Platinic.	Auric.	Mercuric.	Mercurous.	Plumbic.	Arsenic.	Antimonic.	Stannic.	Stannous.	Silver.	Bismuth.	Cupric.	Cadmium.	Ferric.	Aluminum.	Chromic.	Cobaltic.	Nickelic.	Manganous.	Zinc.	Barium.	Strontium.	Calcium.	Magnesium.	Sodium.	Ammonium.	Potassium.	Hydrogen.
Hydrate	A	A	A	A	A	(W)	A–I	A–I	A	A	A	A	A	A	A	A–I	A	A	A	A	(W)	(W)	(W)	A	W	W	W	W
Nitrate	W	W	W	W	W	—	—	W	—	W	W	W	W	W	W	W	W	W	W	W	W	W	W	W	W	W	W	W
Carbonate	A	—	A	A	A	—	—	A	A	A	A	A	A	A	—	A	A	A	A	A	A	A	A	A	W	W	W	W
Acetate	—	—	W	(W)	W	—	—	W	W	(W)	W	W	W	W	W	W	W	W	W	W	W	W	W	W	W	W	W	W
Oxalate	W	—	A	A	A	—	A	W	A	A	A	A	A	A	A	W	A	A	A	A	A	A	A	A	W	W	W	W
Cyanide	(A)	W	W	W	A	—	A	—	—	(A)	A	A	(W)	W	—	A	A	A	A	A	W	W	W	W	W	W	W	W
Chloride	W	W	W	A	(W)	W	W	W	W	I	W	W	W	W	W	W	W	W	W	W	W	W	W	W	W	W	W	W
Bromide	W	W	(W)	A	(W)	W	W	W	W	I	W	W	W	W	W	W	W	W	W	W	W	W	W	W	W	W	W	W
Iodide	A	A	(A)	A	(W)	(W)	A	(W)	(W)	I	A	—	W	W	—	W	W	W	W	W	W	W	W	W	W	W	W	W
Fluoride	W	W	W	A	A	W	W	(W)	W	W	W	(W)	(W)	W	A–I	W	(W)	(W)	A	(W)	A	A	A–I	A	W	W	W	W
Sulphide	A	A	A	A	A	A	A	A	A	A	A	A	A	A	A	A	A	A	A	A	W	W	W	W	W	W	W	W
Sulphite	W	—	A	A	A	—	A	W	(W)	A	A	W	(W)	(W)	A	W	A	A	A	(W)	A	A	W	W	W	W	W	W
Sulphate	W	W	W	W	A–I	—	W	W	W	(W)	W	W	(W)	W	W	W	W	W	W	W	I	I	(W)	W	W	W	W	W
Phosphate	—	A	A	A	A	W	W	A	A	A	A	A	A	A	A	A	A	A	A	A	A	A	A	A	W	W	W	W
Borate	—	—	(W)	—	(W)	—	—	—	(W)	(W)	(W)	(W)	(W)	(W)	(W)	(W)	(W)	W	(W)	(W)	(W)	(W)	(W)	(W)	W	W	W	W
Silicate	—	—	—	—	—	—	—	—	—	—	—	A	—	A	A–I	—	—	—	A	A	A	A	A–I	A–I	W	—	W	(W)I
Arsenite	—	—	A	A	A	—	W	A	A	A	—	A	—	A	—	—	A	A	A	—	(W)	(W)	A	—	W	W	A	W
Chromate	A	—	A	A	A	—	A	A	A	A	A	(W)	A	A	—	A	(W)	W	W	W	A	(W)	(W)	W	W	W	W	W

W, means soluble in water.
(W), difficultly soluble in water.

A, freely soluble in acids.
(A), difficultly soluble in acids.

I, insoluble in water and acids.

TABLE IV.—DERIVATION OF ELEMENTAL NAMES.

	Name.	*Derivation.*	*Discoverer.*
1.	Aluminum.	Latin *alumen*, alum.	Wöhler.
2.	Antimony.	Arabic *al-ithmidun.*	Basil Valentine.
3.	Arsenic.	Greek *arsenicon*, potent.	Schroeder.
4.	Barium.	Greek *barus*, heavy.	Davy.
5.	Bismuth.	German *wisemat.*	Agricola.
6.	Boron.	Arabic *bûraq*, borax.	Davy.
7.	Bromine.	Greek *bromos*, stench.	Balard.
8.	Cadmium.	Greek *cadmia*, calamine.	Stromeyer.
9.	Cæsium.	Latin *cæsius*, blue.	Bunsen.
10.	Calcium.	Latin *calx*, lime.	Davy.
11.	Carbon.	Latin *carbo*, coal.	Ancients.
12.	Cerium.	Latin *ceres*, the planet.	Berzelius and Hisinger.
13.	Chlorine.	Greek *chloros*, green.	Scheele.
14.	Chromium.	Greek *chroma*, color.	Vauquelin.
15.	Cobalt.	German *kobold*, sprite.	Brandt.
16.	Columbium.	*Columbia*, America.	Hatchett.
17.	Copper.	Latin *cyprium*, Cyprus.	Ancients.
18.	Didymium.	Greek *didumos*, twins.	Mosander.
19.	Erbium.	Swedish *Ytterby*.	Mosander.
20.	Fluorine.	Latin *fluo*, to flow.	(Not isolated.)
21.	Gold.		Ancients.
22.	Glucinum.	Greek *glukus*, sweet.	Wöhler.
23.	Hydrogen.	Greek *hudor* and *gennao.*	Cavendish.
24.	Indium.	*Indigo*, a blue dye	Reich and Richter.
25.	Iodine.	Greek *ion*, a violet.	Courtois.
26.	Iron.		Ancients.
27.	Iridium.	Latin *iris*, rainbow.	Tennant.
28.	Lead.		Ancients.
29.	Lanthanum.	Greek *lanthano*, to conceal.	Mosander.
30.	Lithium.	Greek *lithos*, a stone.	Arfvedson.
31.	Magnesium.	*Magnesia*, Asia Minor.	Bussy.
32.	Manganese.	*Magnesia*, Asia Minor.	Gahn.
33.	Mercury.	*Name of planet.*	Ancients.
34.	Molybdenum.	Greek *molybdos*, lead.	Hjelm.
35.	Nickel.	German *kupfernickel.*	Cronstedt.
36.	Nitrogen.	Greek *nitron* and *gennao.*	Rutherford
37.	Osmium.	Greek *osme*, odor.	Tennant.
38.	Oxygen.	Greek *oxus* and *gennao.*	Priestley.
39.	Phosphorus.	Greek *phos* and *phero.*	Brandt.
40.	Platinum.	Spanish *platina.*	Wood.
41.	Potassium.	English *pot-ashes.*	Davy.
42.	Palladium.	*Name of planet.*	Wollaston.
43.	Rhodium.	Greek *rhodon*, a rose.	Wollaston.
44.	Rubidium.	Latin *rubidus*, red.	Bunsen.
45.	Ruthenium.	*Ruthenia*, Russia.	Claus.
46.	Selenium.	Greek *selene*, the moon.	Berzelius.
47.	Silicon.	Latin *silex*, flint.	Berzelius.
48.	Silver.		Ancients.
49.	Sodium.	Latin *salsola*, soda.	Davy.
50.	Strontium.	*Strontian*, Scotland.	Davy.
51.	Sulphur.	Latin *sulfur.*	Ancients.
52.	Tantalum.	*Tantalus*, a deity.	Ekeberg.
53.	Tellurium.	Greek *tellus*, the earth.	Klaproth.
54.	Thallium	Greek *thallos*, a green twig.	Crookes.
55.	Thorium.	*Thor*, a Swedish deity.	Berzelius.
56.	Tin.		Ancients.
57.	Titanium.	*Titans*, myth.	Klaproth.
58.	Tungsten.	Swedish *heavy stone.*	De Luyart.
59.	Uranium.	*Name of planet.*	Klaproth.
60.	Vanadium.	*Vanadis*, deity.	Sefstrom.
61.	Yttrium.	*Ytterby*, Sweden.	Wöhler.
62.	Zinc.		Paracelsus.
63.	Zirconium.	Ceylonese, *zircon.*	Berzelius.

INDEX.

www.ingramcontent.com/pod-product-compliance
Lightning Source LLC
LaVergne TN
LVHW010157110826
845151LV00002B/540

* 9 7 8 1 4 2 5 5 3 6 3 8 1 *